Material Characterization using Electron Holography

Material Characterization using Electron Holography

Daisuke Shindo
Takeshi Tomita

WILEY-VCH

Authors

Dr. Daisuke Shindo
RIKEN
Center for Emergent Matter Science
2-1 Hirosawa, Wako
351-0198 Saitama
Japan

Mr. Takeshi Tomita
242-12 Fuchigami
197-0833 Akiruno-city, Tokyo
Japan

Cover Image: © Daisuke Shindo, Zentaro Akase, Direct observation of electric and magnetic fields of functional materials, Mater. Sci. & Eng. R 142 (2020) 100564, with permission from Elsevier

Library of Congress Card No.: applied for

British Library Cataloguing-in-Publication Data
A catalogue record for this book is available from the British Library.

Bibliographic information published by the Deutsche Nationalbibliothek
The Deutsche Nationalbibliothek lists this publication in the Deutsche Nationalbibliografie; detailed bibliographic data are available on the Internet at <http://dnb.d-nb.de>.

Print ISBN: 978-3-527-34804-6
ePDF ISBN: 978-3-527-82969-9
ePub ISBN: 978-3-527-82970-5
oBook ISBN: 978-3-527-82971-2

Typesetting Straive, Chennai, India
Printing and Binding CPI Group (UK) Ltd, Croydon, CR0 4YY

C107664_031022

Contents

Preface

Transmission electron microscopy (TEM) is widely utilized to clarify the microstructures of various functional materials. To comprehensively understand a material's functionality coupling electric and magnetic properties, it is necessary to characterize the electromagnetic fields in and around the material. Among the various TEM techniques, electron holography is unique in its ability to visualize electric and magnetic fields by utilizing the electron interference effect. This book presents the basis, various applications, and the latest developments of electron holography in four parts.

Part I Introduction emphasizes the importance of the field concept, which is closely related to the theory of relativity. It also briefly explains the basic constitution of transmission electron microscopes along with their mathematical formulation, which is utilized in subsequent parts.

Part II Principles and Practice explains the basic principles of electron holography and the utilization of transmission electron microscopes with special instrumentation. In addition to explaining Maxwell's equations and electromagnetic potentials, it presents the simulation procedures for reconstructing phase images and interference fringes.

Part III Application describes the visualization of electromagnetic fields in and around various functional materials as a means to identify their functional electromagnetic properties. It also describes the clarification and interpretation of the electromagnetic functionalities of a wide variety of materials based on extensive in situ observations of electromagnetic fields created by applying electric or magnetic field at various temperatures.

Part IV Visualization of Collective Electron Motions and Their Interpretation describes the extension of electron holography to the visualization of the collective motions of electrons. Theoretical considerations on the point-charge behavior and interference effect of electrons are presented along with a discussion of the relevant quantum mechanics and the general theory of relativity.

We are grateful to the late Akira Tonomura, the late Shinji Aizawa, Yoshinori Tokura, Toshiaki Tanigaki, Joong J. Kim, Hyun S. Park, Tsuyoshi Matsuda, Xiuzhen Yu, Yasukazu Murakami, Akira Taniyama, Zentaro Akase, Yoichi Ikematsu, Hideyuki Magara, Naoyuki Kawamoto, Yoshimasa A. Ono, Nobuhiko Ohno, Shinichi Ohno, Mitsuru Morita, Hiromitsu Kawase, Kei Hirata,

Takafumi Sato, Akira Yasuhara, Masao Inoue, Yoshitaka Aoyama, Tetsuo Oikawa, Katsushige Tsuno, Gyeong S. Park, Jun-Mo Yang, Young-Gil Park, Jung H. Yoo, Ki H. Kim, Ken Harda, Yoh Iwasaki, Keiko Shimada, Chiari Itonaga, Satoko Takahashi, Kodai Niitsu, Zheng Liu, Weixing Xia, Youhui Gao, Tetsuya Akashi, Hiroto Kasai, Kimi Matsuyama, Hiroyuki Shinada, and Nobuyuki Osakabe for their generous support and invaluable collaboration.

Wako, Japan, 14 December 2021

Daisuke Shindo and Takeshi Tomita
RIKEN

List of Specimens

Part I

Introduction

Material Characterization using Electron Holography, First Edition. Daisuke Shindo and Takeshi Tomita.

1

Importance of Electromagnetic Field and Its Visualization

Transmission electron microscopy (TEM) has been widely utilized to clarify the microstructures of various functional materials. In addition to bright-field and dark-field imaging methods for observing various lattice defects [1–5], high-resolution TEM [6–10] has been used for direct observation of atomic arrangements projected along the incident electron beam direction. Such detailed atomic arrangements can now be observed with a resolution of less than 0.1 nm.

Scanning transmission electron microscopy (STEM) is commonly used for elemental mapping at the atomic level with a microprobe (diameter, less than 0.1 nm) and a beam scanning system [11–14]. Analytical electron microscopy with energy-dispersive X-ray spectroscopy (EDS) and electron energy-loss spectroscopy (EELS) have also been used for composition and electronic state analyses [15–19]. EELS studies can now be performed with an energy resolution of less than 0.1 eV by using monochromators.

To comprehensively understand a material's functionality based on coupling electric and magnetic properties, the electromagnetic fields should be characterized inside and around the material. Among the various TEM techniques, electron holography is unique in its ability to quantitatively visualize electromagnetic fields on the nanometer scale. Here, in relation to visualizing electromagnetic fields, we highlight the importance of the field concept in reference to Einstein and Infeld [20]:

> A new concept appears in physics, the most important invention since Newton's time: the field. It needed great scientific imagination to realize that it is not the charges nor the particles but the field in the space between the charges and particles which is essential for the description of physical phenomena. The field concept proves…
>
> …………
>
> The theory of relativity arises from the field problems…
>
> From "The Evolution of Physics" by A. Einstein and L. Infeld, Cambridge University Press, Cambridge, 2nd ed. (1978) p. 244.

If we bring a magnetic material near a U-shaped magnet, as in Figure 1.1a, we will feel a force between the material and the magnet. This phenomenon can be

Material Characterization using Electron Holography, First Edition. Daisuke Shindo and Takeshi Tomita.

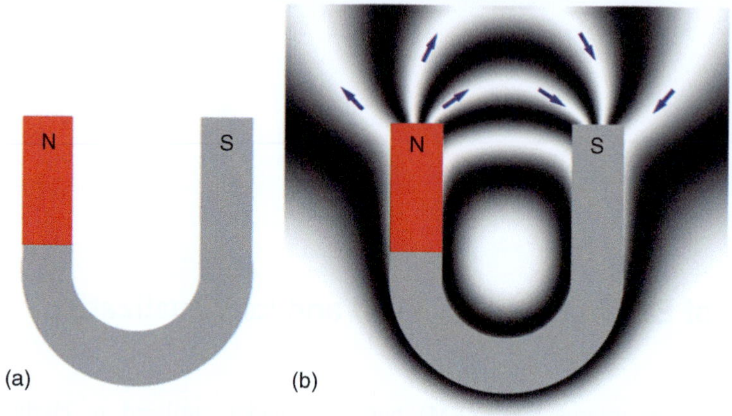

Figure 1.1 (a) U-shaped magnet. (b) Simulated magnetic field around magnet. Arrows indicate the direction of a magnetic flux.

explained by the existence of a magnetic field, which can be simulated as shown in Figure 1.1b.

Though a magnetic field cannot be observed with the naked eye or even with conventional microscopy techniques, it can be visualized using electron holography. Figure 1.2a shows a TEM image of a Co–Zr–O magnetic material [22]. The magnetic information can be recorded by first creating a hologram through the interference effect of incident electrons (Figure 1.2b). Then, by processing the hologram with a Fourier transformation, the magnetic fields both inside and outside the material can be directly visualized (Figure 1.2c). The importance of visualizing electric fields is discussed in Section 6.6.

Electron holography is thus a useful technique for directly visualizing electromagnetic fields. This book addresses the theory and application of electron holography, including the fundamental formulations of electromagnetic fields and relativity. The basic formulations of Maxwell's equations in relation to the special theory of relativity are presented in order to understand the formulation of electromagnetic visualization. The last section of this chapter covers the basic principles of TEM for the specific and detailed explanations of electron microscope hardware in the following chapters.

On the basis of these formulations and explanations, outlines of electron holography and the basic constitution of electron microscopes are explained in the former of Part II. Further detailed explanations of the hardware of transmission electron microscopes for electron holography with special instrumentation are presented in the latter of Part II.

Part III describes the extensive application of electron holography to various functional materials with respect to the principles and instrumentation of electron holography presented in Parts I and II. Techniques for visualizing and interpretating electromagnetic fields in and around materials are introduced for a wide variety

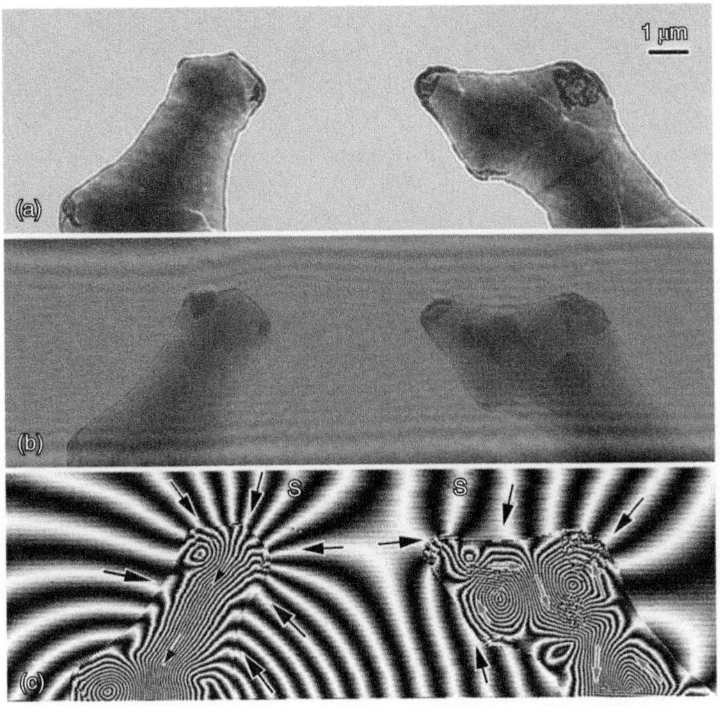

Figure 1.2 (a) TEM image showing two pieces of Co–Zr–O magnetic specimen. (b) Hologram obtained through interference effect of incident electrons. (c) Reconstructed phase image showing detailed magnetic fields both inside and outside the material. Image in (a) was observed under slightly defocused condition. Domain wall contrast (see Section 6.3.1) and absorption contrast due to thickness change appear. Source: Shindo and Akase [21], with permission from ELSEVIER.

of functional materials using a computer simulation. Part III also discusses the effectiveness of using advanced special attachments for in situ observation of electromagnetic fields in order to understand the electromagnetic properties of functional materials.

Part IV focuses on one of the most interesting applications of electron holography to visualize the motion of electrons. The stationary orbital formation and accumulation of secondary electrons around insulating materials, which depend on the material's surface morphology, can be visualized by detecting the fluctuations of the electric fields due to the motions of the secondary electrons. Furthermore, the magnetic flux due to electron spin polarization can be detected by applying an external magnetic field to the secondary electrons. Finally, on the basis of the theory of relativity, electron interference is interpreted using a "spinning linear wave" model proposed by the authors. With this model, the formation of electron interference fringes is successfully reproduced by simulation as a function of the number of incident electrons.

References

1 Hirsch, P.S., Howie, A., Nicholson, R.B. et al. (1977). *Electron Microscopy of Thin Crystals*. New York, FL: Robert E. Krieger Publishing Co. Inc.

2 Amelinckx, D.D.S., Landuyt, J., and Tenderloo, G. (1997). *Handbook of Microscopy*. Weinheim: VCH.

3 Cockayne, D.J.H., Ray, I.L.F., and Whelan, M.J. (1968). *Proceedings. 4 th European.Regional.Conference.Electron Microscopy*, 129. Vaticana, Roma: Tipografia Poliglotta Vaticana.

4 Howie, A. and Basinski, Z.S. (1968). *Philos. Mag. A* 17: 1039.

5 Cockayne, D.J.H., Ray, I.L.F., and Whelan, M.J. (1969). *Philos. Mag. A* 20: 1265.

6 Cowley, J.M. (1990). *Diffraction Physics*. New York: North-Holland.

7 Horiuchi, S. (1994). *Fundamentals of HREM*. Amsterdam: North-Holland.

8 Spence, J.C.H. (1988). *Experimental High Resolution Electron Microscopy*. New York: Oxford University Press.

9 Shindo, D. and Hiraga, K. (1998). *High Resolution Electron Microscopy For Materials Science*. Tokyo: Springer.

10 Spence, J.C.H. (1999). *Mater. Sci. Eng., R* 26: 1.

11 Pennycook, S.J., Berger, S.D., and Culbertson, R.J. (1986). *J. Microsc.* 144: 229.

12 Morishita, S., Ishikawa, R., Kohno, Y. et al. (2018). *Microscopy* 67: 46.

13 Shibata, N., Seki, T., Sánchez-Santolino, G. et al. (2017). *Nat. Commun.* 8: 15631.

14 Shibata, N., Kohno, Y., Nakamura, A. et al. (2019). *Nat. Commun.* 10: 2308.

15 Joy, D.C., Roming, A.D., and Goldstein, J. (1986). *Principles of Analytical Electron Microscopy*. Springer Science & Business Media.

16 Reimer, L. (1984). *Transmission Electron Microscopy: Physics of Image Formation and Microanalysis*. Springer.

17 Egerton, R.F. (1996). *Electron Energy-Loss Spectroscopy in the Electron Microscope*, 2e. New York: Plenum.

18 Williams, D.B. and Carter, C.B. (1996). *Transmission Electron Microscopy: A Textbook For Materials Science*. Springer.

19 Shindo, D. and Oikawa, T. (2002). *Analytical Electron Microscopy for Materials Science*. Tokyo: Springer-Verlag.

20 Einstein, A. and Infeld, L. (1978). *The Evolution of Physics*, 2e. Cambridge: Cambridge University Press.

21 Shindo, D. and Akase, Z. (2020). *Mater. Sci. Eng. R Reports* 142: 100564. https://doi.org/10.1016/j.mser.2020.100564.

22 Liu, Z., Shindo, D., Ohnuma, S., and Fujimori, H. (2003). *J. Magn. Magn. Mater.* 262: 308.

2
Maxwell's Equations and Special Relativity

As the basis for understanding the description of electromagnetic field visualization by using electron holography in Part II, the relationship between electromagnetic fields and electromagnetic potentials is presented through Maxwell's equations (Section 2.1). As a basis for understanding the discussion of the gravitational field in relation to electron coherency in the framework of the general theory of relativity in Part IV, Maxwell's equations are reformulated in the framework of the special theory of relativity (Section 2.2).

2.1 Maxwell's Equations and Electromagnetic Potentials

In this section, we sort out Maxwell's equations and their relationships with electromagnetic potentials. This is important for understanding the principles of electron holography on the basis of scalar and vector potentials discussed in Part II. We also reformulate Maxwell's equations in accordance with the special theory of relativity. This reformulation is fundamentally important to understanding the Lorentz covariance and extending it to the discussion of the general theory of relativity in Section 12.3.

Here, Maxwell's equations with electric charge density ρ and electric current density $\vec{j}$ in SI units (Système International d'Unités) are written as

$$\operatorname{div} \vec{B} = 0 \tag{2.1}$$

$$\frac{\partial \vec{B}}{\partial t} + \operatorname{rot} \vec{E} = 0 \tag{2.2}$$

$$\operatorname{div} \vec{D} = \rho \tag{2.3}$$

$$\operatorname{rot} \vec{H} - \frac{\partial \vec{D}}{\partial t} = \vec{j} \tag{2.4}$$

where $\vec{D}$ and $\vec{E}$ are the electric flux density and electric field, respectively, and $\vec{B}$ and $\vec{H}$ are the magnetic flux density and magnetic field, respectively. Their relationships are expressed as

$$\vec{D} = \varepsilon_0 \vec{E}, \quad \vec{B} = \mu_0 \vec{H} \tag{2.5}$$

Material Characterization using Electron Holography, First Edition. Daisuke Shindo and Takeshi Tomita.

where ε_0 and μ_0 are, respectively, the dielectric constant and magnetic permeability for vacuum:

$$\varepsilon_0 \mu_0 = 1/c^2 \tag{2.6}$$

According to Eqs. (2.1) and (2.2), $\vec{B}$ and $\vec{E}$ are given by scalar and vector potentials φ and $\vec{A}$:

$$\vec{B} = \mathrm{rot}\,\vec{A} \tag{2.7}$$

$$\vec{E} = -\frac{\partial \vec{A}}{\partial t} - \mathrm{grad}\,\varphi \tag{2.8}$$

By inserting Eqs. (2.7) and (2.8) into Eqs. (2.3) and (2.4), we obtain

$$\square\,\varphi = \rho/\varepsilon_0 \tag{2.9}$$

$$\square\,\vec{A} = \mu_0 \vec{j} \tag{2.10}$$

$$\left(\square \equiv \left(\frac{\partial}{c\partial t}\right)^2 - \left(\frac{\partial}{\partial x}\right)^2 - \left(\frac{\partial}{\partial y}\right)^2 - \left(\frac{\partial}{\partial z}\right)^2\right)$$

Equations (2.9) and (2.10) are equivalent to Eqs. (2.3) and (2.4) under the Lorenz condition [1]; i.e.

$$\mathrm{div}\,\vec{A} + \frac{\partial \varphi}{c^2 \partial t} = 0 \tag{2.11}$$

This is because, from Eqs. (2.9) and (2.10),

$$\square\left(\mathrm{div}\,\vec{A} + \frac{\partial \varphi}{c^2 \partial t}\right) = \mu_0 \left(\mathrm{div}\,\vec{j} + \frac{\partial \rho}{\partial t}\right) \tag{2.12}$$

where the right side is zero due to the charge conservation law.

2.2 Maxwell's Equations Formulated Using Special Relativity

On the basis of Maxwell's equations (Eqs. (2.1)–(2.4)), the field around a charged particle at rest, e.g. the electric field, can be evaluated using Eq. (2.3). When the particle or observer moves, the magnetic field formed has to be taken into account using Eq. (2.4).

In the special theory of relativity, the physical laws hold in their simplest forms for any coordinate system subjected to a uniform translational motion relative to another coordinate system (Lorentz covariance). Under this condition and with the same light velocity for both systems, Maxwell's equations are rewritten as follows.

To specify a point in the space–time of physics, we use a four-coordinate representation $x^\mu (\mu = 0, 1, 2, 3)$:

$$x^0 = ct, \quad x^1 = x, \quad x^2 = y, \quad x^3 = z \tag{2.13}$$

A quantity with the suffix written in the upper position is called a "contravariant element."

Let us take a point close to the point x^μ, and let the coordinates of the point be $x+dx^\mu$. The four coordinates forming the displacement are considered to be components of a vector, a "contravariant vector." In the theory of special relativity, the squared magnitude of line element ds with light velocity in accordance with Dirac's notations [2]

$$(ds)^2 = (dx^0)^2 - (dx^1)^2 - (dx^2)^2 - (dx^3)^2 \tag{2.14}$$

is independent of the orientation, and the local coordinate system and is called a "scalar." For generalization purposes, this equation is rewritten as

$$(ds)^2 = \sum_{\mu=0}^{3} \sum_{\nu=0}^{3} \eta_{\mu\nu} dx^\mu dx^\nu \tag{2.15}$$

where $\eta_{\mu\nu}$ is the Minkowskian metric tensor:

$$\eta_{\mu\nu} = \eta_{\nu\mu} = \left\{ \begin{array}{c} 1\ (\mu = \nu = 0) \\ 0\ (\mu \neq \nu) \\ -1\ (\mu = \nu = 1, 2, 3) \end{array} \right\} \tag{2.16}$$

Since the tensor has the suffix written in the lower position, it is called a "covariant tensor" (of the 2nd rank). Following Einstein [3], we omit the summation symbol, so Eq. (2.15) is simply written as

$$(ds)^2 = \eta_{\mu\nu}\, dx^\mu\, dx^\nu \tag{2.17}$$

In the following, when we have a term with the same suffix in both the upper and lower positions, we consider it to be a summation taken over four values, from 0 to 3.

Consider the contravariant vector A^μ related to the scalar and vector potentials above:

$$A^0 = \varphi/c, \quad A^1 = A_x, \quad A^2 = A_y, \quad A^3 = A_z \tag{2.18}$$

Contravariant vector A^μ is transformed to covariant vector A_μ using $\eta_{\mu\nu}$ and

$$A_\mu \equiv \eta_{\mu\nu} A^\nu \tag{2.19}$$

Thus,

$$A_0 = A^0 = \varphi/c, \quad A_k = -A^k\ (k = 1, 2, 3) \tag{2.20}$$

Similarly, a four-current j_μ consisting of charge density (ρ) and current (j_x, j_y, j_z) is given as

$$j^0 = c\rho, \quad j^1 = j_x, \quad j^2 = j_y, \quad j^3 = j_z \tag{2.21}$$

With these notations, two of Maxwell's equations ((2.9) and (2.10)) are written as

$$\square A^\lambda = \mu_0 j^{\lambda\,(\lambda = 0,1,2,3)} \tag{2.22}$$

$$\left(\square \equiv \eta^{\mu\nu} \frac{\partial}{\partial x^\mu} \frac{\partial}{\partial x^\nu} = \left(\frac{\partial}{c\partial t}\right)^2 - \left(\frac{\partial}{\partial x}\right)^2 - \left(\frac{\partial}{\partial y}\right)^2 - \left(\frac{\partial}{\partial z}\right)^2 \right)$$

$$\left(\eta^{\mu\nu} = \eta^{\nu\mu} = \begin{cases} 1\ (\mu = \nu = 0) \\ 0\ (\mu \neq \nu) \\ -1\ (\mu = \nu = 1, 2, 3) \end{cases} \right)$$

The Lorenz condition [1] is written as

$$\partial_\mu A^\mu = 0 \tag{2.23}$$

$$\left(\partial_\mu = \frac{\partial}{\partial x^\mu}\right)$$

To represent electric and magnetic fields in the framework of the special theory of relativity, we introduce an antisymmetric covariant tensor of the 2nd rank; i.e.

$$f_{\mu\nu} = -f_{\nu\mu} = \frac{\partial A_\nu}{\partial x^\mu} - \frac{\partial A_\mu}{\partial x^\nu} = \partial_\mu A_\nu - \partial_\nu A_\mu \tag{2.24}$$

This gives us the correspondence between $f_{\mu\nu}$ and $(\vec{E}, \vec{B})$:

$$\begin{pmatrix} f_{00} & f_{01} & f_{02} & f_{03} \\ f_{10} & f_{11} & f_{12} & f_{13} \\ f_{20} & f_{21} & f_{22} & f_{23} \\ f_{30} & f_{31} & f_{32} & f_{33} \end{pmatrix} = \begin{pmatrix} 0 & E_x/c & E_y/c & E_z/c \\ -E_x/c & 0 & -B_z & B_y \\ -E_y/c & B_z & 0 & -B_x \\ -E_z/c & -B_y & B_x & 0 \end{pmatrix} \tag{2.25}$$

Thus, the first two of Maxwell's equations (Eqs. (2.1) and (2.2)) are written as

$$\partial_\lambda f_{\mu\nu} + \partial_\mu f_{\nu\lambda} + \partial_\nu f_{\lambda\mu} = 0 \tag{2.26}$$

The second two of Maxwell's equations (Eqs. (2.3) and (2.4)) are written as

$$\partial_\nu f^{\lambda\nu} = \mu_0 j^\lambda \tag{2.27}$$

These equations formulated in the special theory of relativity hold in the same form for any coordinate system subjected to a uniform translational motion relative to another coordinate system (Lorentz covariance).

Given these reformulations of Maxwell's equations in accordance with the special theory of relativity, we can express Maxwell's energy–momentum tensor $T^{\mu\nu}$ as

$$T^{\mu\nu} = \frac{1}{\mu_0}\left(\eta_{\lambda\sigma} f^{\lambda\mu} f^{\sigma\nu} - \frac{1}{4}\eta^{\mu\nu} f^{\alpha\beta} f_{\alpha\beta}\right) = T^{\nu\mu} \tag{2.28}$$

This formulation is used for the discussion in Section 12.3.

References

1 Lorenz, L.V. (1867). *Phil. Mag. Ser. 3* 34: 287. https://doi.org/10.1080/14786446708639882.
2 Dirac, P.A.M. (1975). *Geneal Theory of Relativity*. Wiley.
3 Einstein, A. (1916). *Ann. der Phys. Ser. 4* 49: 769.

3

Basis of Transmission Electron Microscopy

In this chapter, we formulate the scattering of incident electrons. This is important to understand the basis of transmission electron microscopy (TEM) and for mathematical treatments of electron holography data.

The plane wave of an incident electron is generally given by $\exp[i(\vec{k} \cdot \vec{r} - \omega t)]$, where k and ω are the wave number and angular frequency given by $2\pi/\lambda$ and $2\pi\nu$ (λ: wavelength, ν: frequency), respectively. For a specimen in the stationary condition, the phase and amplitude changes of the plane wave are independent of time. Thus, the plane wave of an incident electron is simply given by $\exp(i\vec{k} \cdot \vec{r})$. If the effect of the specimen on the plane wave is given by $q(x, y)$, the scattering amplitude at the observation point (s, t) shown in Figure 3.1a is given by

$$\psi(s,t) = C \iint q(x,y) \frac{\exp(ikr)}{r}\, dx\, dy \tag{3.1}$$

where C is a constant and r is given as

$$r = \sqrt{R^2 + (x-s)^2 + (y-t)^2} \tag{3.2}$$

When the first term in the square root is much larger than the sum of the second and third terms, the binomial theorem is applied to r:

$$\begin{aligned} r &= R\sqrt{1 + \frac{(x-s)^2 + (y-t)^2}{R^2}} \\ &= R + \frac{(x-s)^2 + (y-t)^2}{2R^2} - \frac{[(x-s)^2 + (y-t)^2]^2}{8R^3} \\ &= R + \frac{s^2+t^2}{2R} - \frac{sx+ty}{R} + \frac{x^2+y^2}{2R} - \frac{[(x-s)^2 + (y-t)^2]^2}{8R^3} \end{aligned} \tag{3.3}$$

The terms necessary in Eq. (3.3) depend on the size of R. Inclusion of up to the third term is referred to as "Fraunhofer approximation," while that up to fourth term is "Fresnel approximation." The approximation to use is set by comparing the value of $[k\,(x^2+y^2)/2R]$ with $\pi/2$. For example, if the specimen area is an L-square, $[k\,(x^2+y^2)/2R]$ has a value of $\pi/2$ for $R = L^2/\lambda$, and thus

$$R > \frac{L^2}{\lambda} \quad \text{(area for Fraunhofer approximation)} \tag{3.4}$$

$$R < \frac{L^2}{\lambda} \quad \text{(area for Fresnel approximation)} \tag{3.5}$$

Material Characterization using Electron Holography, First Edition. Daisuke Shindo and Takeshi Tomita.

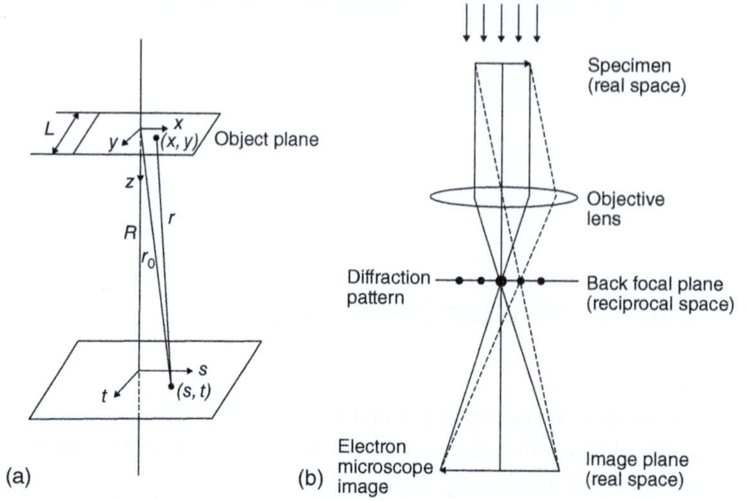

Figure 3.1 (a) Schematic showing electron scattering in a specimen. (b) Diagram of TEM imaging process.

For the Fraunhofer approximation, Eq. (3.3) is given by

$$r \simeq r_0 - sx/r_0 - ty/r_0 \tag{3.6}$$

Thus, Eq. (3.1) is written as

$$\psi(u, v) = C' \iint q(x, y) \exp[-2\pi i\,(ux + vy)] dx\,dy \tag{3.7}$$

where C', u, and v are given, respectively, as

$$C' = C\ \exp(ik\,r_0)/r_0, \quad u = s/\lambda r_0, \quad v = t/\lambda r_0 \tag{3.8}$$

Since the right-hand side of Eq. (3.7) is a form of the Fourier transformation, the scattering amplitude $\psi(u, v)$ can be considered given by the Fourier transformation [1, 2]. The conditions for Fraunhofer diffraction, as given by Eq. (3.7), are satisfied for electron diffraction with an electron lens system in a transmission electron microscope, as in Figure 3.1b. In the following chapters, mathematical treatment using Fourier transformation is frequently utilized in processes such as scattering from a specimen to the back focal plane and image formation through the back focal plane onto the image plane.

References

1 Cowley, J.M. (1975). *Diffraction Physics*. Amsterdam: North-Holland.
2 Shindo, D. and Hiraga, K. (1998). *High Resolution Electron Microscopy For Materials Science*. Tokyo: Springer.

Part II

Principles and Practice

Material Characterization using Electron Holography, First Edition. Daisuke Shindo and Takeshi Tomita.

4

Principles of Electron Holography

In this chapter, we first explain two types of electron holography and then describe the visualization process of electromagnetic fields. Finally, we outline the computer simulation of reconstructed phase images of the electromagnetic field.

4.1 Types of Electron Holography

The concept of holography was introduced by Gabor in 1948 [1]. Using an interference microscope and a method for reconstructing wavefronts, Gabor aimed to improve the resolution of electron microscopes by recording the amplitude distribution of interference fringes resulting from the interaction of an object and a coherent reference wave.

He used in-line holography, which is the simplest way of producing a hologram. An example of geometrical configurations of in-line holography using a transmission electron microscope is illustrated in Figure 4.1a. The transmitted plane wave acts as a reference wave. The object wave indicated by dark region and the reference wave propagate parallel to the optical axis. However, at the time, this approach was problematic because, first, the spatial or lateral coherence length (Section 5.3) was insufficient to form a hologram covering a sufficiently wide region and, second, incoherent scattering effects added an unwanted strong background. In addition, two geometrical limitations included the specimen size, which should be small compared with the region of the reference wave at the specimen position (object plane), and the overlapping of the reconstructed image with a defocused conjugate image due to formation of both the reconstructed image and its conjugate image along the same optical axis. Various improvements have since been made and applied to overcome these limitations.

The geometrical limitations of in-line holography led to the invention of off-axis electron holography in which a biprism was introduced (Figure 4.1b). This causes the object wave (dark region) passing through the specimen to separate from the reference wave. This off-axis electron holography was invented and extensively developed by Möllendstedt [2] and his colleagues at Tübingen University.

Off-axis electron holography has become widely utilized and has been used to characterize materials by using a highly coherent electron beam emitted from

Material Characterization using Electron Holography, First Edition. Daisuke Shindo and Takeshi Tomita.

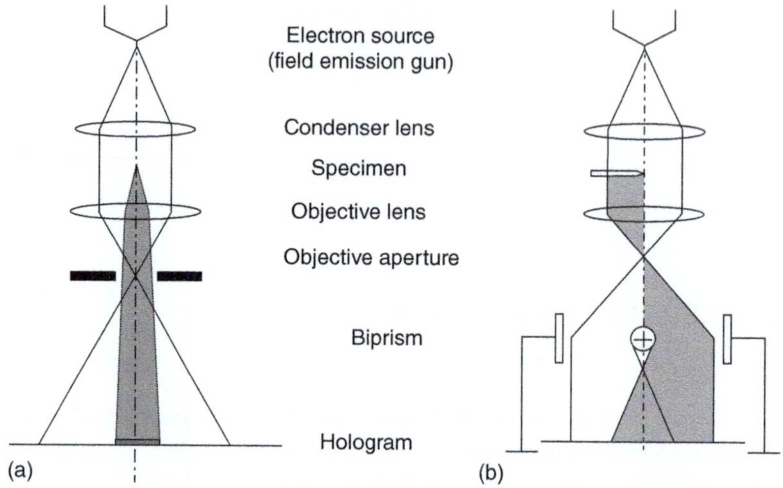

Figure 4.1 Two types of electron holography: (a) in-line electron holography; (b) off-axis electron holography.

a field emission gun (Section 5.1) and various specimen preparation techniques (Section 6.7) for suppressing the inelastic scattering effects. Hereinafter, the explanations of electron holography focus on this latter method ("off-axis electron holography" is referred to as "electron holography").

4.2 Outline of Electron Holography

Figure 4.2 illustrates the basic geometric configuration of electron holography (strictly speaking, off-axis electron holography) and its imaging process (right panel) [3–11]. Electron holography has a two-step imaging process. First, a hologram is formed with a biprism in a transmission electron microscope, whereby an object wave passing through a specimen interferes with a reference wave passing through the vacuum. Next, the phase shift of the electrons is extracted from the hologram by using an optical reconstruction system or a computer for executing a Fourier transformation. The latter approach using digital data treatment has become widely used due to the availability of quantitative software.

A highly coherent electron beam is emitted from a field emission gun (Table 4.1) and collimated to illuminate a thin specimen through a condenser lens system. Compared with thermionic emission electron guns, field emission ones have high brightness, a small source size, and small energy spread. However, they are relatively expensive and require careful treatment.

When the voltage of the biprism is zero (Figure 4.3a), a conventional transmission electron microscopy (TEM) image of the absorption–diffraction contrast [12] is obtained. When the voltage is increased, interference fringes form (Figure 4.3b). When the voltage is further increased, the interference fringes cover the specimen region to record the electric and magnetic information (Figure 4.3c).

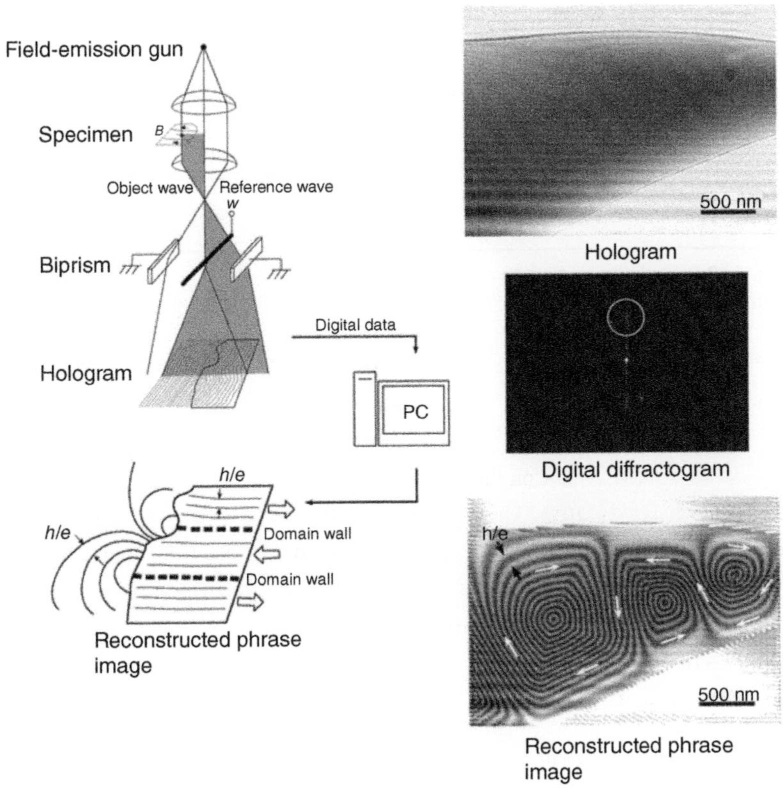

Figure 4.2 Illustration showing geometric configuration of electron holography and its imaging process on right.

Care should be taken when analyzing magnetic materials because the magnetic field at the specimen position in a transmission electron microscope is more than 1 T, which is strong enough to destroy or modify the inherent magnetic domain structure. Thus, various specialized instruments should be utilized (Section 6.4). The specimens for TEM observation should be thin, so their domain structure differs greatly from that in bulk materials. The use of a high-voltage electron microscope is thus effective (Section 6.4).

In the formulation, we assume that the object is illuminated by a plane wave of a unit amplitude having angular frequency (ω) and wave number (k) parallel to the optical axis, i.e. $\exp[i(kz - \omega t)]$. The change of the scattering amplitude of the plane wave due to the object is, in general, described as $q(x, y)$:

$$q(x,y) = a(x,y)\exp\{i\phi(x,y)\} \tag{4.1}$$

where $a(x, y)$ and $\phi(x, y)$ are real functions describing the amplitude change and phase shift due to the electromagnetic field, respectively [12]. The amplitude function $a(x, y)$ is attributed to absorption and/or diffraction in the specimen. The phase

Table 4.1 Characteristics of various 200-kV electron guns.

			Field emission		
	Thermionic emission		**Thermal field emission gun (FEG)**		**Cold FEG**
Characteristic	**W**	**LaB_6**	**$ZrO/W^{(100)}$**	**$W^{(100)}$**	**$W^{(310)}$**
Brightness at 200 kV (A $(cm^2 \cdot sr)^{-1}$)	$\sim 5 \times 10^5$	$\sim 5 \times 10^6$	$\sim 5 \times 10^8$	$\sim 5 \times 10^8$	$\sim 1 \times 10^9$
Source size	50 μm	10 μm	0.2~1 μm	~100 nm	~100 nm
Energy spread (eV)	2.3	1.5	0.6~0.8	0.6~0.8	0.3~0.5
			Conditions for usage		
Pressure (Pa)	10^{-3}	10^{-5}	10^{-7}	10^{-7}	10^{-8}
Temperature (K)	2800	1800	1800	1600	300
			Emission		
Current (μA)	~100	~20	~100	20~30	10~30
Stability for short time	1%	1%	1%	5%	5%
Stability for long time	1%/h	3%/h	1%/h	6%/h	5%/15 min
Current efficiency	30%	10%	10%	10%	1%
Maintenance	No need	No need	Need some times for setup	Build up several times for a new tip	Flashing necessary every few hours
Price	Cheap	Cheap	Expensive	Expensive	Expensive
Operation	Easy, simple	Easy, simple	A little compli-cate	A little delicate	A little delicate

shift $\phi(x, y)$ is expressed as

$$\phi(\mathrm{x}, \mathrm{y}) = \sigma \int \varphi(x, y, z)dz - \frac{\mathrm{e}}{\hbar} \int \vec{A}(x, y, z)d\vec{z} \tag{4.2}$$

where $\varphi(x, y, z)$ and $\vec{A}(x, y, z)$ represent the electric (scalar) and magnetic vector potentials, respectively, and σ is the interaction constant, which is determined from velocity v of the incident electrons:

$$\sigma = \frac{e}{\hbar v} = \frac{2\pi}{\lambda V(1 + \sqrt{1 - (v/c)^2})} \tag{4.3}$$

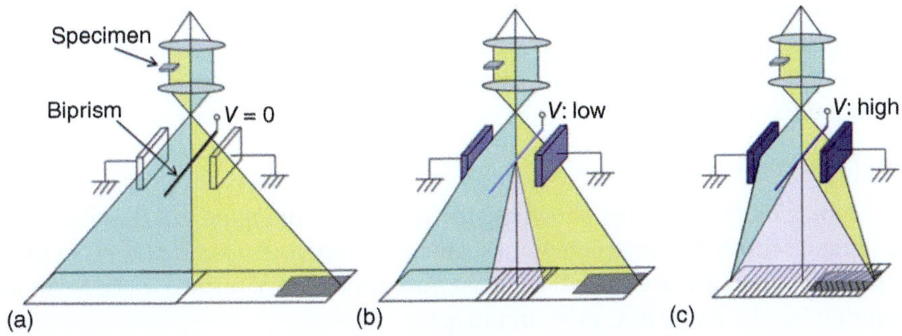

Figure 4.3 Formation of interference fringes with increasing biprism voltage V: (a) zero, (b) low, (c) high.

where λ and V are the wavelength and accelerating voltage of the incident electrons, respectively, and c is the speed of light [13, 14]. The values of σ as a function of the accelerating voltage are listed in Appendix A.

In conventional TEM (illustrated in Figure 4.3a), the intensity (squared absolute value of $q(x,y)$) is given by

$$I(x, y) = |q(x, y)|^2 = a^2(x, y) \tag{4.4}$$

which indicates that only the absorption–diffraction contrast is determined. This means that the phase information, consisting of the electric and magnetic vector potentials, cannot be obtained.

With electron holography (illustrated in Figure 4.3c), a hologram $I_{\text{hol}}(x, y)$ is formed using the interference between the object wave and the reference wave (see also Figure 1.2b):

$$\begin{aligned} I_{\text{hol}}(x, y) &= \left| a(x, y) \exp\left(-\pi i \frac{\alpha_h}{\lambda} x + i\phi(x, y)\right) + \exp\left(\pi i \frac{\alpha_h}{\lambda} x\right) \right|^2 \\ &= 1 + a^2(x, y) + 2a(x, y) \cos\left(\frac{2\pi\alpha_h x}{\lambda} - \phi(x, y)\right) \end{aligned} \tag{4.5}$$

where α_h is a superimposed angle between the object wave and the reference wave. The last term in Eq. (4.5) corresponds to the interference fringes. This equation is based on the assumption of fully coherent and aberration-free conditions.

By Fourier transformation (F) of the hologram, we obtain

$$\begin{aligned} F[I_{\text{hol}}(x, y)] = \delta(x, y) &+ F[a^2(x, y)] \\ &+ F[a(x, y) \exp(\mathrm{i}\phi(x, y)] * \delta(u + \alpha_h/\lambda, v) \\ &+ F[a(x, y) \exp(-\mathrm{i}\phi(x, y)] * \delta(u - \alpha_h/\lambda, v) \end{aligned} \tag{4.6}$$

where * indicates the convolution operation. Information about the phase shift and amplitude change is carried in the sidebands, which correspond to the third and fourth terms, on the right-hand side. By selecting the third term, shifting it by α_h/λ, and performing inverse Fourier transformation (F^{-1}) on this term, we obtain

$$F^{-1}\left[F\left[a(x, y) \exp(i\phi(x, y))\right] * \delta(u, v)\right] = a(x, y) \exp(i\phi(x, y)) \tag{4.7}$$

Thus, we obtain both the phase shift and the amplitude change in the reconstructed phase image. In most of the experimental results presented in this book, the intensity of the reconstructed phase image $I_{\mathrm{ph}}(x, y)$ is given in terms of the cosine function of $\phi(x, y)$:

$$I_{\mathrm{ph}}(x, y) = \cos(\phi(x, y)) \tag{4.8}$$

These imaging processes are shown on the right panel of Figure 4.2. It is also possible to represent the reconstructed phase image in terms of $1 + \cos(\phi(x, y))$ so that only positive values are obtained. If necessary, amplification of the phase can be done by multiplying the phase $\phi(x, y)$ by an integer n.

4.3 Comparison of Phase Shifts Due to Scalar and Vector Potentials

4.3.1 Phase Shift Due to Scalar Potential

When a specimen is a non-magnetic material, the phase shift is due to a scalar potential. We evaluate the phase shift using Eq. (4.2),

$$\phi(x, y) = \sigma \int \varphi(x, y, z) dz \tag{4.9}$$

As shown in Figure 4.4a, when we have a thin specimen with thickness t, we compare the phase shift of the incident electron wave at positions C and D under the specimen. From Eq. (4.9), if the specimen is homogeneous and its inner potential is assumed to be constant, we obtain

$$\begin{aligned} \phi(x_C, y_C) - \phi(x_D, y_D) &= \sigma \int_0^t \varphi(x_C, y_C, z) - \varphi(x_D, y_D, z) dz \\ &= 0 \end{aligned} \tag{4.10}$$

Thus, for such a specimen with constant thickness, the phase shift remains the same, so information about the inner potential cannot be obtained.

4.3.2 Phase Shift Due to Vector Potential

Now we consider the phase shift due to a magnetic field. If the specimen thickness is uniform in the viewing field as shown in Figure 4.4b, $\phi(x, y)$ can be expressed by

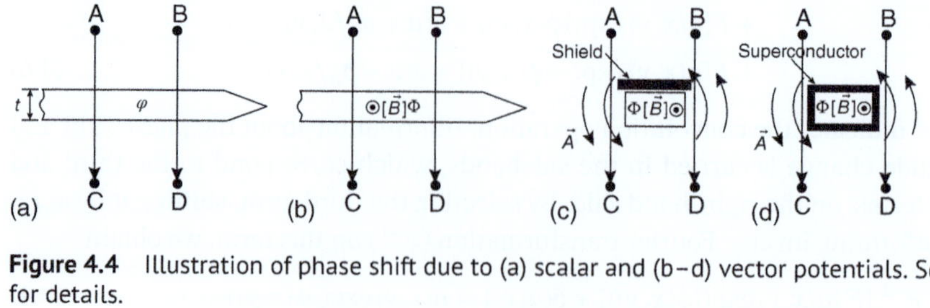

Figure 4.4 Illustration of phase shift due to (a) scalar and (b–d) vector potentials. See text for details.

vector potential $\vec{A}$ related to magnetic flux density B:

$$\phi(x,y) = -\frac{e}{\hbar}\int \vec{A}(x,y,z)d\vec{z} \tag{4.11}$$

where e and $\hbar$ stand for the elementary electric charge and Planck's constant divided by 2π, respectively. Assuming that the electron wave has the same phase at points A and B (above the specimen) in Figure 4.4b, the phase difference between points C and D (below the specimen) is given by

$$\begin{aligned}\phi(x_C,y_C) - \phi(x_D,y_D) &= -\frac{e}{\hbar}\int_A^C \vec{A}(x_C,y_C,z)\cdot d\vec{z} + \frac{e}{\hbar}\int_B^D \vec{A}(x_D,y_D,z)\cdot d\vec{z} \\ &= \frac{e}{\hbar}\oint_{ABDC} \vec{A}(x,y,s)d\vec{s}\end{aligned} \tag{4.12}$$

Using the vector potential in Eq. (2.7) and using Stokes' theorem, we obtain

$$\begin{aligned}\oint \vec{A}\cdot d\vec{s} &= \iint \mathrm{rot}\vec{A}\cdot d\vec{S} \\ &= \iint \vec{B}\cdot d\vec{S} \\ &= \iint B_n \; dS\end{aligned} \tag{4.13}$$

where B_n is the component of $\vec{B}$ normal to area dS. Thus, we get

$$\begin{aligned}\phi(x_C,y_C) - \phi(x_D,y_D) &= \frac{e}{\hbar}\iint B_n \, dS \\ &= \frac{e}{\hbar}\Phi\end{aligned} \tag{4.14}$$

where Φ is the magnetic flux passing through and normal to the area $ABDC$. If the phase difference between $\phi(x_C, y_C)$ and $\phi(x_D, y_D)$ is 2π, we have

$$\frac{e}{\hbar}\Phi - 2\pi \tag{4.15}$$

and then

$$\Phi = \frac{h}{e} = 4.1\times 10^{-15}\ \mathrm{Wb} \tag{4.16}$$

This value is twice the flux quantum ($h/2e$). Thus, we have the relationship between the magnetic flux inside the specimen and width l of the white or black lines in the reconstructed phase image ($\cos(\phi(x, y))$), which corresponds to a 2π phase difference:

$$\Phi = l\,t\,B_n = h/e \tag{4.17}$$

where t is the specimen thickness. From Eq. (4.17), the relation between l and B_n is given by

$$l = \frac{h}{etB_n} \tag{4.18}$$

Different from the scalar potential, the phase shift appears as shown in Eq. (4.14) for a constant specimen thickness. Furthermore, even if the electron beam does not pass through the specimen or its magnetic field $\vec{B}$, there is a difference in phase shifts between positions C and D because $\vec{B}$ exists between two beam paths: A–C and B–D. This difference results from the existence of vector potential $\vec{A}$, shown in Figure 4.4c. This is the Aharonov–Bohm effect [15].

Tonomura et al. extended the electron holography system using a toroidal ferromagnet covered with a superconductor (Figure 4.4d). Using this extended system, they were able to clearly prove the physical reality of vector potentials [16].

4.3.3 Effect of Thickness Change on Phase Shifts Due to Scalar and Vector Potentials

In the above descriptions, we assumed the specimen had a constant thickness (Figure 4.4). In general, the thickness of a specimen prepared by ion milling and/or focused ion beam systems is assumed to change, especially in the edge region. Figure 4.5a shows a hologram of an as-quenched specimen of Fe–Cu–Ni–Si–B prepared by ion milling. The interference fringes are deformed in the specimen region, while in the vacuum region outside the specimen, the interference fringes are straight. In the right and left parts of the specimen region, the interference fringes are deformed with convex-downward shapes as indicated by arrowheads, while they are deformed with convex-upward shapes in the lower central part. These differences are attributed to the magnetic flux distribution of closure domains (Figure 4.5b).

Simulation of reconstructed phase images was carried out to take into account the wedge shape of the specimen (Figure 4.6a). The inner potential and the saturation magnetization for the specimen were assumed to be 17 V and 1.2 T, respectively. The direction of the magnetic flux was assumed to be parallel to the specimen edge corresponding to the closure domain of a soft magnetic material. The effect of the stray magnetic field was neglected. The reconstruction of the phase image in Figure 4.6b was simulated assuming the inner potential only without a magnetic field, while the reconstruction of the phase image in (c) was simulated assuming a magnetic field only. Figure 4.6d,e shows reconstructed phase images for which the contributions of both the inner potential and magnetic field were included. The density of the black or white bands, which corresponds to the change in phase shift along the direction perpendicular to the specimen edge, differed depending on the direction of the magnetic flux. Regions P and R in Figure 4.5b correspond to the simulation result shown in Figure 4.6d, while region Q corresponds to that in Figure 4.6e.

4.3.4 Electric Information

The phase information of the electric and magnetic fields can be separated using the "time-reversal operation of an electron beam" proposed by Tonomura et al. [17]. In this process, a pair of electron holograms are obtained for the same area, but the foil is reversed against the incident electron beam: the sign (positive or negative) of the phase depends on the direction of the incident electron beam for the magnetic field, whereas it is independent of the electric field (Figure 4.7). Other techniques for

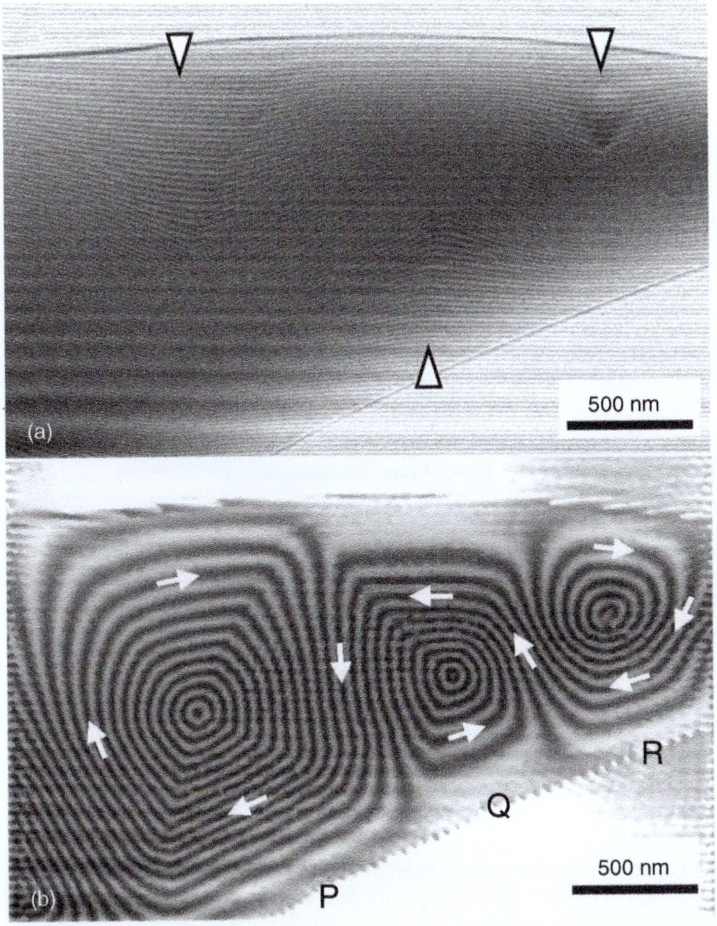

Figure 4.5 (a) Hologram and (b) reconstructed phase image of quenched $Fe_{73.5}Cu_1Nb_3Si_{13.5}B_9$ specimen. Arrows in (b) indicate direction of magnetic flux.

separating the phase information of the electric field from that of the magnetic field include applying an external magnetic field and setting the temperature to various values around the magnetic phase transformation temperature (Curie temperature). These techniques are described in Chapter 8.

4.4 Analysis of Reconstructed Phase Images by Computer Simulation

Computer simulation using Maxwell's equations can be performed to analyze electric and magnetic fields quantitatively. For example, to clarify the magnetization in detail from the observed magnetic flux distribution, simulation can be performed in accordance with the flow chart shown in Figure 4.8 [18].

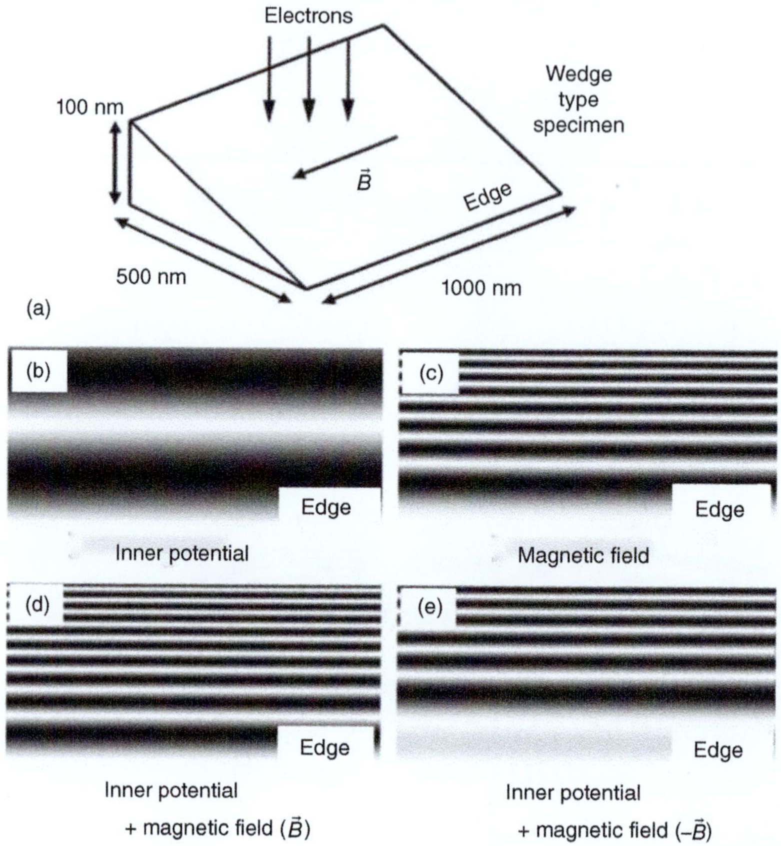

Figure 4.6 (a) Model specimen used for simulation of phase image reconstruction. (b) Reconstructed phase image of electric field only (inner potential: 17 V). (c) Reconstructed phase image of magnetic field only (magnetic flux: 1.2 T). (d) Simulated image obtained from (b) and (c). (e) Simulated image obtained from (b) and magnetic field with same magnitude but in opposite direction to field in (c).

Following the flow chart, Aoyama et al. [18] simulated the reconstruction of phase images of an Nd–Fe–B-based nanocomposite magnet consisting of nanometer-size grains of hard and soft magnetic phases [19, 20]. Although the specimen was prepared by ion milling, the observation was carried out for an area distant from the specimen edge, so the specimen thickness change was neglected. They considered a model structure in which a spherical grain of the hard magnetic phase is surrounded by a matrix (i.e. a mixture of hard and soft magnetic phases), as shown in the center of Figure 4.9b, where θ represents the angle between the magnetization vector and the foil plane. As seen in the figure, the features of the reconstructed phase images changed dramatically as the magnetization vector in the spherical hard magnetic phase was tilted off the foil plane. By comparing the simulation results with observations, such as that in (a), it is possible to estimate the tilt angles of the magnetization vectors in hard magnetic grains dispersed in the specimen. This means

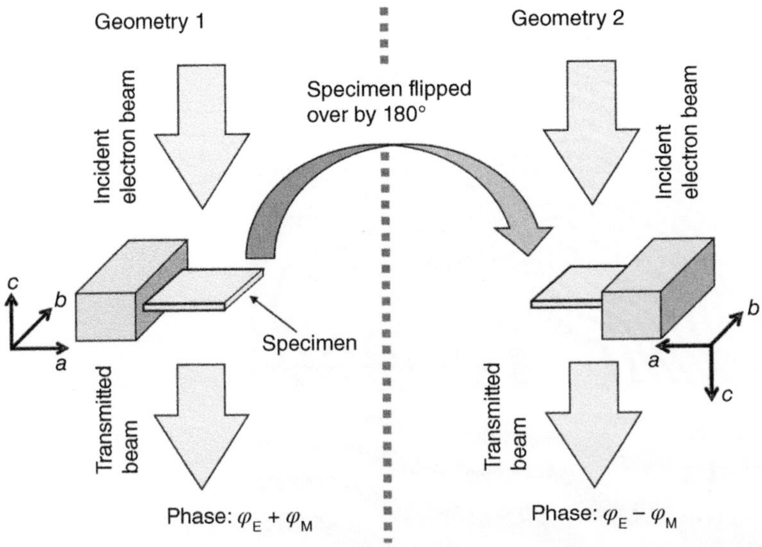

Figure 4.7 Diagram illustrating process of separating electric and magnetic fields. Sign of magnetic contribution is reversed when incident beam is inverted with respect to specimen.

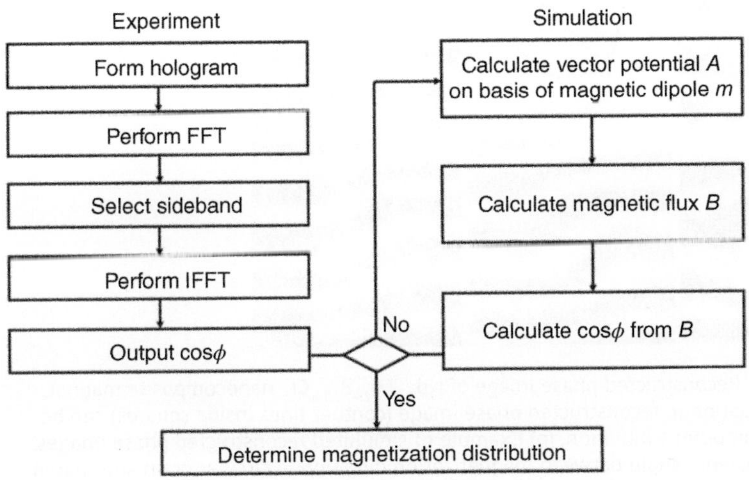

Figure 4.8 Flow chart for determining magnetization distribution by electron holography. FFT, fast Fourier transformation; IFFT, inverse fast Fourier transformation.

that, although, in principle, electron holography reveals the in-plane component of the magnetic flux, computer simulation can provide useful information about ferromagnetic domains for which the magnetization vectors are tilted off the foil plane. An explanation of three-dimensional magnetic structure analysis using a dual-axis 360°-rotation specimen holder is presented in Section 8.4.

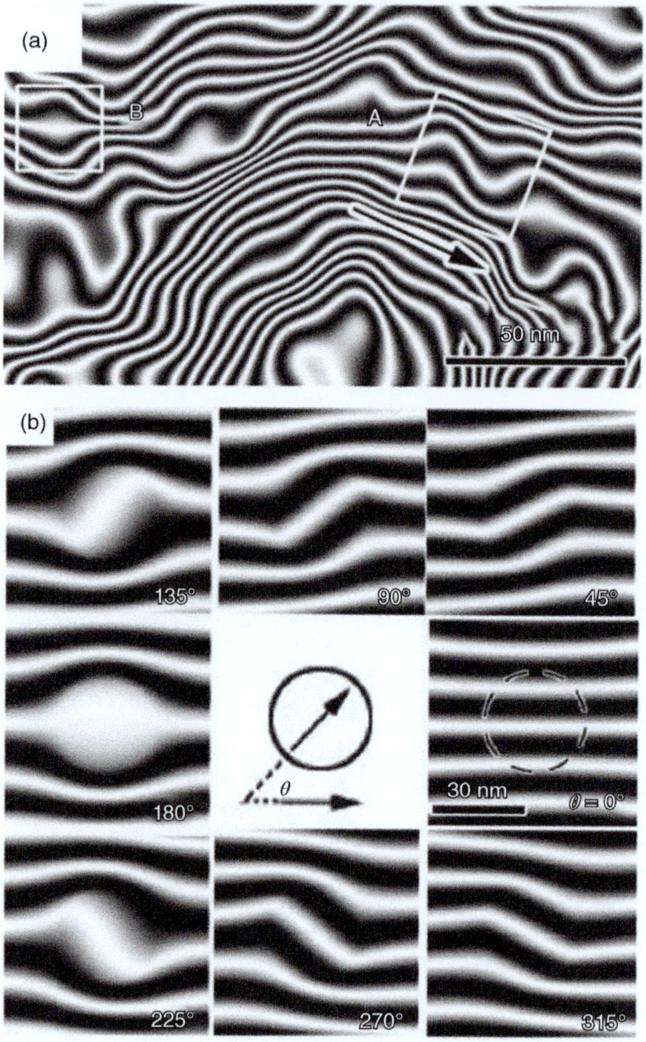

Figure 4.9 (a) Reconstructed phase image of $Nd_{4.5}Fe_{74}B_{18.5}Cr_3$ nanocomposite magnet. Source of fluctuation in reconstructed phase image (contour lines inside squares) can be identified by computer simulation. (b) Example of simulated reconstructed phase images. Variable θ represents angle between magnetization direction in circular grain and that in matrix, as shown by schematic at center.

References

1 Gabor, D. (1948). *Nature* 161: 777.

2 Möllenstedt, G. (1999). The history of the electron biprism. In: *Introduction to Electron Holography* (ed. E. Völkl, L.F. Allard and D.C. Joy), 1. New York: Kluwer Academic.

3 Shindo, D. and Oikawa, T. (2002). *Analytical Electron Microscopy for Materials Science*. Tokyo: Springer-Verlag.

4 Tonomura, A., Allard, L.F., Pozzi, G. et al. (1995). *Electron Holography, Proceedings of the International Workshop on Electron Holography, Holiday Inn World's Fair, Knoxville, Tennessee, USA, August 29–31, 1994*. Elsevier.

5 Tonomura, A. (1999). *Electron Holography*, 2e. Springer-Verlag.

6 Lichte, H. and Lehmann, M. (2007). *Rep. Prog. Phys.* 71: 16102.

7 Shindo, D. and Murakami, Y. (2011). *Microscopy* 60: S225.

8 Völkl, E., Allard, L.F., and Joy, D.C. (2013). *Introduction to Electron Holography*. Springer Science & Business Media.

9 Tanigaki, T., Harada, K., Murakami, Y. et al. (2016). *J. Phys. D: Appl. Phys.* 49: 244001.

10 Zweck, J. (2016). *J. Phys. Condens. Matter* 28: 403001.

11 Dunin-Borkowski, R.E., Kovács, A., Kasama, T. et al. (2019). Electron holography. In *Springer Handbook of Microscopy* (ed. P.W. Hawkes and J.C.H. Spence), 767–818. Cham: Springer International Publishing.

12 Shindo, D. and Hiraga, K. (1998). *High Resolution Electron Microscopy For Materials Science*. Tokyo: Springer.

13 Lichte, H. (2008). *Ultramicroscopy* 108: 256.

14 Shindo, D. and Murakami, Y. (2008). *J. Phys. D: Appl. Phys.* 41: 183002.

15 Aharonov, Y. and Bohm, D. (1959). Significance of electromagnetic potentials in the quantum theory. *Phys. Rev.* 115 (3): 485.

16 Tonomura, A., Osakabe, N., Matsuda, T. et al. (1986). *Phys. Rev. Lett.* 56 (8): 792.

17 Tonomura, A., Matsuda, T., Endo, J. et al. (1986). *Phys. Rev. B: Condens. Matter* 34: 3397.

18 Aoyama, Y., Park, Y.-G., and Shindo, D. (2005). *Microscopy* 54: 279.

19 Shindo, D. (2025). *Mater. Trans.* 2003: 44.

20 Shindo, D., Park, Y.-G., Murakami, Y. et al. (2003). *Scr. Mater.* 48: 851.

5

Microscope Constitution and Hologram Formation

In this chapter, we explain the basic constitution of a transmission electron microscope for electron holography. On the basis of the explanation of each component, we note the concept of the coherence length and principles of interference fringe formation with simulation.

5.1 Basic Constitution of Transmission Electron Microscope

First, the main components of a transmission electron microscope are described. Then, the appropriate operation of each component is explained. As shown in Figure 5.1 and described below, a standard transmission electron microscope consists of four parts:

(1) Electron gun system
(2) Illumination system
(3) Imaging system
(4) Observation system

Outlines of these systems and their functions are as follows.

5.1.1 Electron Gun System

An electron beam is generated by an electron gun at the top of the electron optical system and accelerated. There are various types of electron guns, as detailed in Table 4.1. For electron holography experiments, a highly coherent electron beam is necessary. As discussed in Section 5.3, the energy spread is directly related to temporal coherence, and the spatial coherence is determined by the illumination angle. For electron holography study, a field emission gun (FEG), either a thermal or cold type, is usually used.

Electrons in metals pass through a potential barrier due to the tunneling effect and can be emitted from the surface of metals because the potential barrier of the metal–vacuum boundary thins when a strong electrical field is applied to the surface of the metal. A cathode fabricated with a sharply pointed shape with a 0.1-μm

Material Characterization using Electron Holography, First Edition. Daisuke Shindo and Takeshi Tomita.

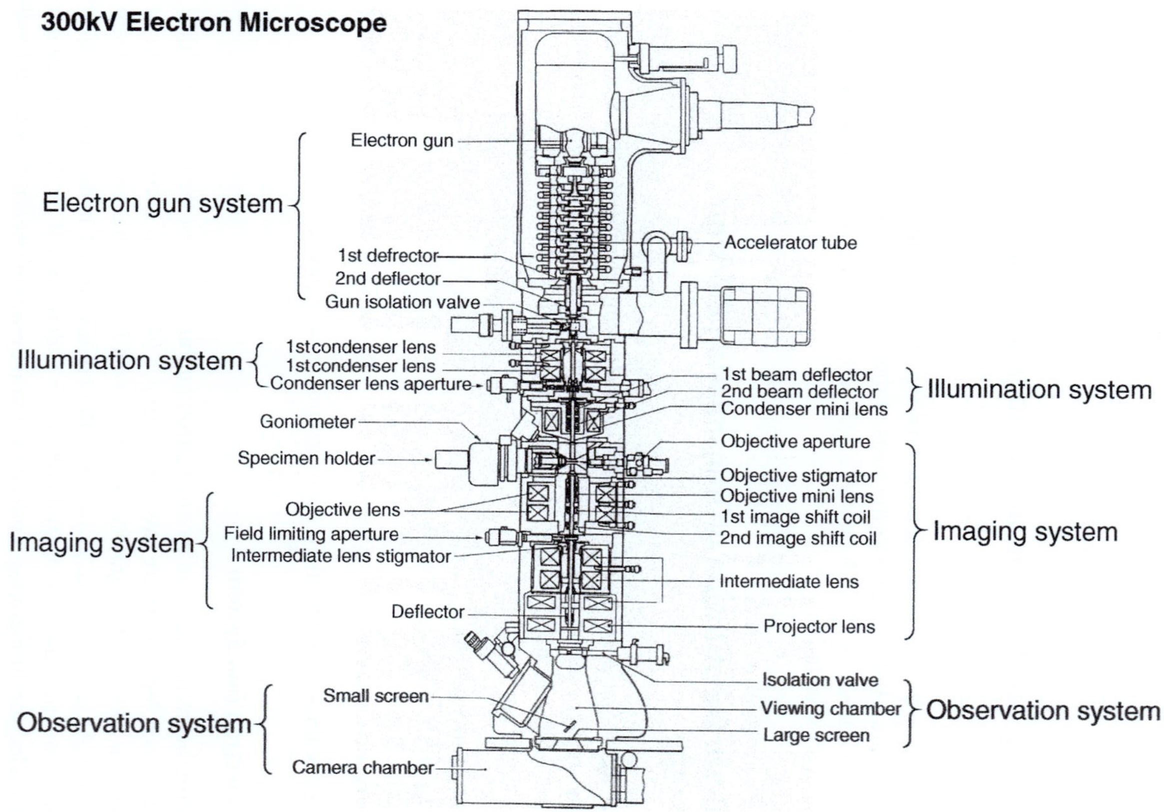

Figure 5.1 Cross section of column in transmission electron microscope (JEM-3000F). A standard transmission electron microscope basically consists of an electron gun system, an illumination system, an imaging system, and an observation system.

radius of curvature to localize the electric field is called an "emitter" or "tip." An FEG produces about 100 times higher electron brightness than a thermionic emission gun made with a LaB_6 single-crystal tip and provides an extremely small electron source. Owing to these characteristics, it is easy to make a small probe and derive high brightness; hence, FEGs are now widely utilized in analytical electron microscopy. The high coherence of the electrons produced with an FEG makes electron holography study possible (see Section 5.3).

A cold FEG uses W with the surface of the (310) plane as an emitter. The emitter works at room temperature without heating. Because the energy spread is as small as 0.3–0.5 eV, it can obtain high-energy resolution in electron energy-loss spectroscopy (EELS). However, residual gas contamination on the emitter surface generates emission noise and/or emission instability. The emission current decreases slowly with an increase in the thickness of the contamination layer. A regular maintenance procedure called "flashing" is thus necessary.

If a ZrO/W tip is used for the electron source, electrons are emitted from the tip with [100]-orientation by heating the tip to about 1800 K and applying an electric potential of about 3 kV between the tip and the first anode located immediately below the tip. The ambient pressure around the tip is kept at about 10^{-8} Pa order so that a stable total emission current of up to 180 μA can be continuously supplied. Only a small fraction (1% or less) of the electrons passes through the first anode aperture, after which they are accelerated by the following electrodes.

With a thermal FEG, the emitter is heated to a lower temperature than the thermal electron emission temperature of 1600–1800 K under a strong electric field, causing electrons to be emitted. The electrons pass through a potential barrier that is reduced by application of the electric field. This phenomenon is called the "Schottky effect." It has the disadvantage of a large energy spread (0.6–0.8 eV) due to heating the emitter compared with that of a cold FEG. On the other hand, it has less emission noise and provides a stable emission current without flashing because there is no adsorption of contamination on the emitter.

With all types of electron guns, the emitted electrons are commonly accelerated with a following acceleration tube. An acceleration tube is indispensable for achieving a long-life tip in a high-voltage gun since large high-voltage discharges often destroy pointed tips. A 300-kV gun has 10–11 accelerator stages, whereas a 200-kV gun has 6–7 stages. To control these high-voltage electron source systems with high stability, it is necessary to have an ultrahigh vacuum and avoid micro-discharges.

5.1.2 Illumination System

The shape and intensity of the beam are controlled in the illumination system. The electron beam is aligned with the optical axis, and the beam diameter is adjusted to cover the field of view in accordance with the purpose of the investigation.

The illumination lens system and the deflector provide functions that cause the electrons to converge on the specimen. An analytical electron microscope can also perform analyses of a small area (≤1 nm in diameter). This is made possible by the ability of the illumination lens system to produce a small probe and by changing the illuminating condition from parallel to convergent.

Parallel illumination on a wide area of a specimen provides highly coherent electron illumination. This illumination is achieved by strongly exciting the condenser mini lens (CM lens) so that the electrons are focused on the prefocal point of the objective prefield. When the CM lens is turned off, the electrons are focused on the specimen by the objective prefield. When the illumination angle α_i becomes large, the electron intensity increases. This condition is suitable for analyzing a small area. Under this condition, a small-diameter probe with relatively high illumination coherence is used. This is suitable for observing convergent-beam electron diffraction (CBED) patterns.

An electron deflector is used for beam alignment, beam tilting, beam shifting, beam scanning, and so on. The deflector, composed of a pair of deflection coils, as shown in Figure 5.2, is called a "double-deflection system" and provides ease of operation. A beam angle of θ_2 relative to the beam axis is achieved by first tilting the beam θ_1 away from the axis with the DEF1 coil and then tilting it back with the DEF2 coil.

The geometrical relationship between θ_1 and θ_2 is given by

$$\tan\theta_2 = l_1/l_2 \tan\theta_1 \tag{5.1}$$

where l_1 is the distance between DEF1 and DEF2 and l_2 is the distance between DEF2 and the specimen. By presetting the current ratio for DEF1 and DEF2 with the balance adjuster, tilt angle θ_2 can be adjusted using only one volume controller. In this system, the beam position on the specimen is kept the same, whereas the tilt angle can be adjusted. Adjusting the deflector balance enables the beam tilt (or beam

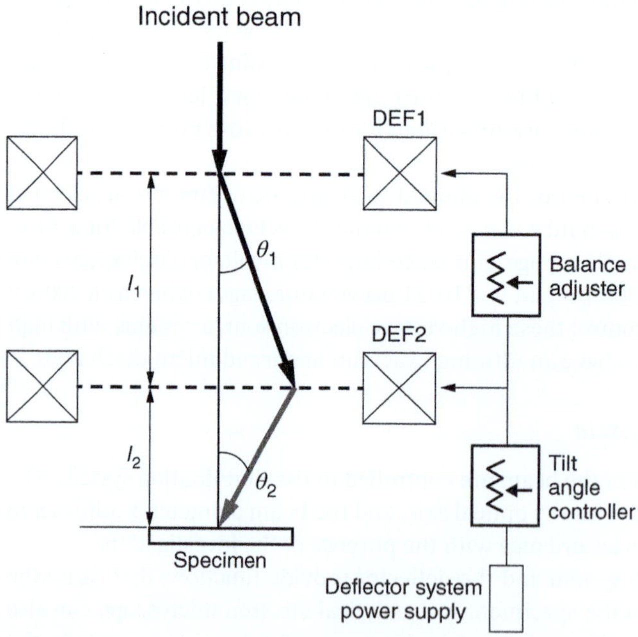

Figure 5.2 Principle of double-deflection system for beam tilt. DEF1: first-stage deflection coil. DEF2: second-stage deflection coil.

shift) to be controlled with only one knob, independent of the beam shift operation (or beam tilt operation). Deflectors are used not only in the illumination system but also in the electron gun, the image-forming system, and the projection system for beam alignment. In the imaging system, the electron beam is transmitted and diffracted by the specimen. The transmitted and diffracted electrons form images and diffraction patterns. A biprism attachment is installed in the objective lens and in the intermediate lens as explained in Section 5.2.

5.1.3 Imaging System

The objective lens forms an image when electrons pass through the specimen. The image quality of a transmission electron microscope is determined mainly by the performance of the objective lens. It is composed of lens coils, a magnetic circuit (yoke), and a pole piece, as shown Figure 5.3. The shape of the pole piece determines the optical properties of the objective lens. Figure 5.4 shows a cross section of the pole piece of the objective lens in a conventional transmission electron microscope. A strong magnetic flux is generated in the space between the upper and lower pole pieces. The specimen is set between the upper and lower pole pieces, with the objective aperture located below the specimen. The objective stigmator is installed below the lower pole piece. The four parameters discussed below are typical, reflecting the main optical properties of the objective lens:

a. Focal length
b. Spherical aberration coefficient
c. Chromatic aberration coefficient
d. Minimum step of defocus

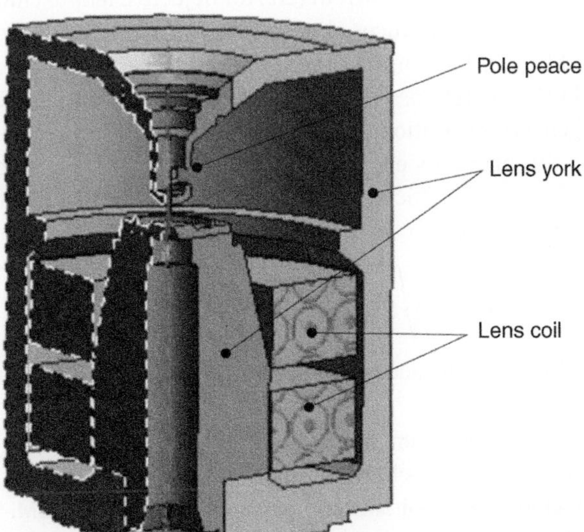

Figure 5.3 Quarter cross section of objective lens system comprising lens coils, magnetic circuit (yoke), and pole piece.

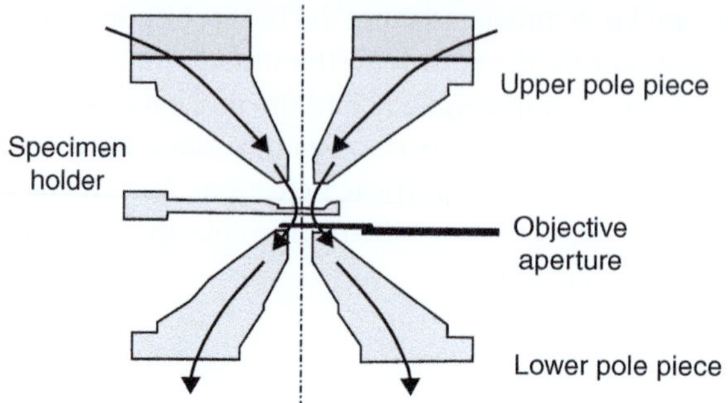

Figure 5.4 Cross section of standard objective pole piece showing position of specimen holder and aperture. Arrows indicated flows of magnetic flux. A strong magnetic flux is generated in space between upper and lower pole pieces.

5.1.3.1 Focal Length

In a magnetic lens, stronger lens excitation generates a shorter focal length. The focal length (f_0) of the objective lens is almost constant because the objective lens is used under conditions of constant lens excitation. In general, a short focal length provides small spherical aberration and high resolution, while a longer focal length provides higher image contrast.

5.1.3.2 Spherical Aberration Coefficient

The point resolution (d) of a transmission electron microscope is determined by the spherical aberration coefficient (Cs) of the objective lens and the wavelength (λ) of the incident electrons:

$$d = 0.65(Cs\,\lambda^3)^{1/4} \tag{5.2}$$

In general, stronger excitation of the objective lens provides a smaller Cs, so it provides a higher point resolution.

5.1.3.3 Chromatic Aberration Coefficient

A variation in the current (ΔI) for objective lens excitation causes the focal length to vary, which produces chromatic aberration. A variation in the accelerating voltage (ΔV_v), in the energy spread of the electrons (ΔV_e) emitted from the emitter, and in the energy loss of the electrons (ΔV_s) due to inelastic scattering in a specimen generates variations in the wavelength, resulting in chromatic aberration.

Stronger lens excitation of the objective lens also provides a smaller chromatic aberration coefficient (C_c). Variation in the focal length (Δ) due to chromatic aberration caused by these factors is given by

$$\Delta = Cc\sqrt{\left(2\frac{\Delta I}{I}\right)^2 + \left(\frac{\Delta V_\text{v}}{V}\right)^2 + \left(\frac{\Delta V_\text{e}}{V}\right)^2 + \left(\frac{\Delta V_\text{s}}{V}\right)^2} \tag{5.3}$$

where V is the accelerating voltage and I is the objective lens excitation current.

5.1.3.4 Minimum Step of Defocus

A defocusing technique for making the objective focus slightly under-focused is widely used to obtain higher image contrast in TEM. Especially in high-resolution electron microscopy, a through-focus method (serially defocusing) is used. With this method, the lower the minimum defocusing step (Δf) is set, the smaller the defocusing series can be used for obtaining through-focus images.

The function of the objective lens forming an image results from the postfield of the lens at the backside of the specimen. On the other hand, a prefield works as a condenser lens in current analytical electron microscopes. Because of this strong prefield, a small electron beam probe can be used. The objective lens of this type, with simultaneous functions of the condenser lens and the imaging lens, is called the condenser–objective (C–O) lens.

The objective aperture, which is generally located at the back focal plane, is used to obtain image contrast. It removes some of the scattered electrons (diffracted waves), resulting in the appearance of the image contrast. This contrast looks like absorption of the incident electrons inside the specimen. The image contrast formed by this mechanism is called "absorption–diffraction contrast" or "amplitude contrast." Under tilted illumination conditions created using deflectors, dark-field imaging can be carried out by selecting a specific wave diffracted by the objective aperture. The back focal plane of the objective lens corresponds to the reciprocal space. The objective aperture can limit some higher spatial frequencies, so it should be carefully used in high-resolution electron microscopy. The image contrast in high-resolution electron microscopy is created by interference between the transmitted wave and diffracted wave, which is called "phase contrast" [1].

Principle of Electron Lens Action The magnetic flux generated by the lens coil is condensed at the tip of a pole piece by the yoke. The pole piece has rotational symmetry around the optical axis, with bore diameter b and gap distance S between the poles, as shown in Figure 5.5. It is designed such that the flux condenses at the gap. The electrons, which pass along the optical axis exactly, do not suffer the Lorentz force due to the magnetic field. An incident electron at distance r from the axis suffers the Lorentz force in the direction from back to front of the paper by an r-component B_{1r} of magnetic field B_1.

The Lorentz force is given by $F = -e(\vec{v} \times \vec{B})$. The direction of the force is given by Fleming's left-hand rule. An electron starts to rotate clockwise but then experiences the r-direction force of z-component B_{2z} of magnetic field B_2, which causes the electron to converge to the optical axis. The electron is thus focused at a point on the axis (focal point). If the rotational motion is neglected, the focusing action of the magnetic lens can be considered to be equivalent to a convex lens of light. The rotation angle is almost proportional to the total magnetic flux of the lens. Consequently, the image formed by the magnetic lens is, in general, a rotated image, not the inverted image obtained with an optical lens.

5.1.4 Observation System

In the observation system, magnified images or diffraction patterns are projected on a screen or a camera system. Projected electron microscopic images and diffraction patterns can be observed on a fluorescent screen in a viewing chamber. Binoculars are usually placed in the viewing chamber to facilitate image focusing.

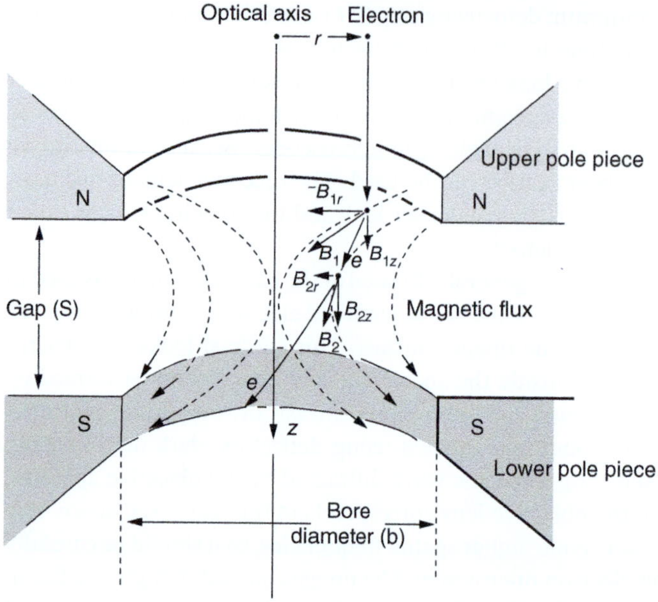

Figure 5.5 Electron path in pole pieces, showing principle of action in electron lens.

The fluorescent screen is an aluminum plate coated with phosphor powder. Lead glass is used for the viewing window to shield the hard X-rays generated in the column of a transmission electron microscope. The thickness of the lead glass must be increased if the accelerating voltage is increased; therefore, observing fine contrast images on the screen is generally difficult in high-voltage electron microscopy. A television (TV) camera installed below the camera chamber is useful in such cases. Television and slow-scan charge-coupled device (CCD) cameras are widely used recording systems for TEM.

5.1.4.1 Television Camera

Using a television (TV) system to observe and record electron microscope images is convenient for in situ observation, multiuser observation, computer input, and instant printing. TV observation of electron microscope images is done using a TV camera combined with a dedicated fluorescent screen. Each image is converted into a light image by using a transmission-type fluorescent screen; it is then transferred to a recording device through a fiber-optic plate or a close-up lens. The resolution of the fluorescent screen is about 100 μm depending on the kind of phosphor, the phosphor coating method, and the accelerating voltage. The fiber-optic plate is composed of many 6-μm fibers. There is little loss of light intensity, and little image distortion is generated in the fiber-optic plate. However, there is occasional blemishing of the fiber and contact patterns of the fibers. Moreover, the small area of the detector limits the field of view of the TV camera. Although the magnification function of a microlens can be used to increase the field of view, doing so results in light absorption and image distortion. An image intensifier in the TV camera can be used to increase its sensitivity.

A CCD does not create shading. Moreover, it is compact and inexpensive. However, it has low sensitivity and a narrow dynamic range for electron intensity.

A video signal generated by a TV camera can be sent to a video signal processor for image processing (i.e. contrast enhancement, shading correction, image accumulation). Image subtraction [(image intensity) – (electron intensity without specimen)] is especially useful for suppressing various blemishes produced in the system. It is easy to transfer a TV image to a computer. A video capture board with an RS-170 or NTSC input terminal can be installed in most personal computers, and a TV can be connected to the video terminal on the board with a video cable.

5.1.4.2 Slow-Scan Charge-Coupled Device Camera

Figure 5.6 shows the configuration of a slow-scan CCD camera [2] used in transmission electron microscopes. Incident electrons are converted into light by an yttrium aluminum garnet (YAG) scintillator and transferred to the CCD through a fiber-optic plate. The light is detected and converted into an electron charge that is temporarily stored in each channel of the semiconductor electrode on the surface of the CCD. The accumulated electrical charge is sequentially transferred to the neighboring pixel while sweeping and then sent from a terminal as an electrical signal. Thus, by storing the electrical charge for some time and reading it out using scanning, a slow-scan CCD camera has higher sensitivity and a wider dynamic range than a real-time CCD camera. The dark current producing the noise and the background on the image can be reduced by cooling the CCD. The digital image data can be displayed on a monitor within seconds after electron exposure in a slow-scan CCD camera.

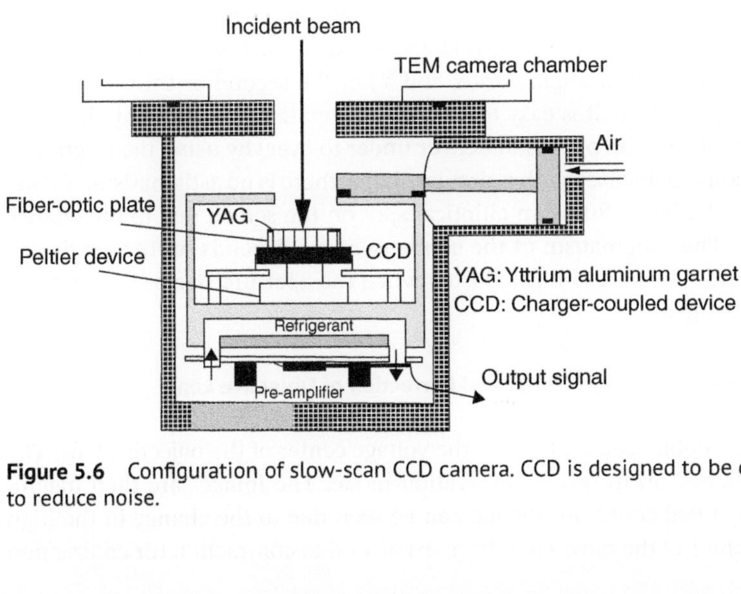

Figure 5.6 Configuration of slow-scan CCD camera. CCD is designed to be cooled in order to reduce noise.

5.1.5 Operation of Transmission Electron Microscope

Lens alignment and astigmatism correction are important adjustments in the operation of a transmission electron microscope.

The microscope adjustments in daily operation are performed in order, starting from the top of the microscope. During the adjustment procedure, the objective excitation current or voltage (objective focus) should be set at the original value ("focus current" or "focus voltage"), and the specimen position should be set at the original Z-position. If, at the start of the procedure, some of the parts deviate greatly from these two conditions, adjusting one part accurately may be difficult. This is overcome by roughly adjusting all the parts and then repeating the adjustments for all parts until finally performing accurate adjustments.

5.1.5.1 Adjustment of Electron Gun

With a thermal FEG, the electric field between the first and second anodes acts as an electrostatic lens, and the axis of this lens needs to be adjusted. This is done by first switching on an anode wobbler and adjusting the alignment of the axis using the gun tilt and shift knobs (X and Y) for gun alignment to stop the movement of the beam in the center.

5.1.5.2 Alignment and Astigmatism Correction of Condenser Lenses

The alignment and astigmatism correction of the condenser lenses are done by first selecting a large spot size (first condenser lens weakly excited) by using the spot size knob and then centering the beam on the screen by using the gun alignment shift knobs (X and Y). Next, a small spot size (first condenser lens strongly excited) is selected by using the spot size knob and then centering the beam on the screen by using the condenser alignment shift knobs (X and Y). These steps are repeated until the beam is centered on the screen without a shift, irrespective of the spot size.

Next, the beam is focused on the screen by using the brightness knob (focusing in the second or third condenser lens). The shape of the focused beam is made circular by adjusting the condenser stigmator (X and Y) of the second or third condenser lens. With this procedure, it is easy to observe astigmatism by changing the focus of the condenser lens from over to under (or under to over) by using the brightness knob. If the beam is defocused with a circular shape, there is no astigmatism. If there is astigmatism, the beam forms an elliptical spot on the screen due to the change in beam focus. The astigmatism of the condenser lens depends on the condenser aperture size and spot size, so it is adjusted when the aperture size or the spot size (or both) are changed.

5.1.5.3 Alignment of Voltage Center and Correction of Objective Lens Astigmatism

A high-voltage wobbler is used to align the voltage center of the objective lens. The wobbler is switched on in image observation mode. The images are then magnified and demagnified continuously and can be seen due to the change in the high voltage. The center of the movement from expansion to contraction (or contraction to expansion) is the high-voltage center. The high-voltage center is centered on the screen by operating the deflectors (tilt X and Y of condenser alignment). The objective aperture is retracted from the optical axis during this operation and reinserted afterward. The use of an image with a triple point in the carbon microgrid facilitates this process. The wobbler center is centered on the screen by using the tilt knobs of the double-deflection system. The tilt knobs of the double deflector system can be used to adjust the wobbler center to the center of the screen as the image has only the movement from expansion to contraction (or contraction to expansion) without the shift. The location of the high-voltage center depends on the spot size and magnification, so it is adjusted when these conditions change. The dependence is especially strong for electron microscopes with a C–O lens producing a strong magnetic prefield. Current center alignment is another way to align the objective lens. The objective wobbler is used to change the objective lens current.

Astigmatism of the objective lens is corrected by using the inserted objective aperture. Because astigmatism depends on the aperture size and position, it must be corrected each time the conditions change. Astigmatism correction is carried out in the low and middle magnification range. The circular or square hole in a specimen (a carbon microgrid is convenient) is used for the adjustment at magnifications lower than ×100 000. In the just-focus condition, the objective stigmator (X and Y) is adjusted to obtain the same image contrast of the hole for two directions perpendicular to each other at an edge. Then, in the under-focus condition, the image contrast and fringe width at the hole edge in both directions are checked to ensure that they are the same. Next, again in the just-focus condition, the stigmator is again adjusted as described above. Finally, in the over-focus condition, the over-focused fringes are checked to ensure that they are the same as for the under-focus condition. This process may be repeated several times.

For magnifications higher than ×200 000, an amorphous region in the specimen or a carbon-contaminated region at the edge of the specimen is located. The thin amorphous carbon film in such regions is suitable for astigmatism correction. First, the objective stigmator (X and Y) is adjusted to remove a unidirectional pattern in the phase contrast image. Next, the objective focus is made under-focused, and the granular contrast image is checked to ensure that a unidirectional pattern does not appear. Then, in the just-focus condition, the stigmator is adjusted as described above. Next, in the over-focus condition, the granular image contrast without a unidirectional pattern is checked to ensure that it is the same as that in the under-focus condition. If necessary, these procedures are repeated several times. Finally, the image contrast without a unidirectional pattern in the just-focus condition is checked to ensure that it is minimum.

5.1.5.4 Correction of Intermediate Lens Astigmatism

Astigmatism of the intermediate lens is corrected by first removing the objective aperture and inserting a selected area aperture. Then, an electron diffraction pattern is observed by controlling a first intermediate lens by using the "DIFF-focus" knob. The intermediate lens stigmator (X and Y) is then adjusted until the shape of the electron beam is circular. For this operation, it is easy to observe the magnitude of the astigmatism by changing the diffraction focus from over to under (or under to over).

5.1.5.5 Alignment of Projector Lens

The project lens is aligned by projecting a diffraction pattern and adjusting the projector lens deflector (X and Y) to center the pattern on the screen. Misalignment of the projector lens does not have a major effect on image quality, so deflection can be used for the purpose of moving a specific part of projected image to be investigated to the center of the screen. Adjusting the image center for an off-axis TV camera is an example application of this procedure.

5.1.5.6 Adjustment of Objective Lens Focus

Adjustment of Z-Position At the low or middle magnification range, when a specimen is tilted, the image of the specimen generally tends to move and escape from the field of view. To suppress such movement of the image by tilting, the axis of the goniometer should be controlled for the specimen position with the following procedures.

First, the specimen is tilted around the tilting axis of a goniometer that has a knob for moving the specimen in the optical axis (Z-axis). After focusing the specimen image, the specimen is tilted by about $\pm 10°$ with the goniometer to check whether the specimen image moves or not. If the image of specimen moves, the specimen positioning knob is used to bring the specimen image to the original position of the screen. At this position, the specimen image is focused again by tilting in the opposite direction. If the image of specimen moves, the specimen positioning knob is used to bring the specimen image to the original position again. These operations are repeated for several times to get the optimum condition.

Adjustment at Low and Middle Magnifications Several basic points must be kept in mind for electron microscope observation. For example, the specimen position should be set exactly at the Z-position, and the objective lens excitation should be set to the just-focus condition. Focus adjustment using the image wobbler is also a basic operation.

First, the objective lens excitation current is set to the just-focus condition. The image of the specimen is then observed, and the image wobbler is switched on. If image wobbling is observed, the specimen position deviates from the exact Z-position. The specimen position (height) is adjusted to stop the wobbling.

If image wobbling is observed when the specimen is at the exact Z-position, the objective lens current may be deviating from the optimum condition. The lens excitation is adjusted with the focus knob to stop the image wobbling. It is difficult to distinguish over-focusing from under-focusing by using image wobbling alone. It may be necessary to change the focus or the Z-position while carefully observing the change in image movement to identify the just-focus condition.

Adjustment at Higher Magnifications At magnifications higher than ×200 000, the image contrast or Fresnel fringes at a specimen edge can be used for focus adjustment. The dark image contrast of Fresnel fringes appears in the over-focus condition, and a fringe with bright image contrast appears in the under-focus condition. In the just-focus condition, the fringes at the edge disappear, and image contrast in a thin area becomes minimum. For high-resolution electron microscope observations that are slightly under-focused, the Scherzer focus is widely used as the optimum focus condition.

5.2 Biprism System

An electron biprism system is installed between the objective lens and intermediate lens as shown Figure 5.7. The configuration of a typical biprism system is shown in Figure 5.8. Outline of biprism wire holder and power connection to electron microscope column are shown in Figure 5.8a. Screws in a biprism wire holder can be used to adjust wire position relative to optical axis as shown in (b). The wire direction relative to a specimen can be adjusted by rotating the biprism wire shown in (c). When a voltage is applied to the wire, the object wave and reference wave are interfered, and fringes are observed on the screen or camera monitor, as shown in Figure 5.9. The ray paths in Figure 5.10 indicates three kinds of magnified hologram imaging processes. Three different magnification modes can be achieved with the lens system consisting of the objective lens (OL), the objective mini lens (OM), and the intermediate

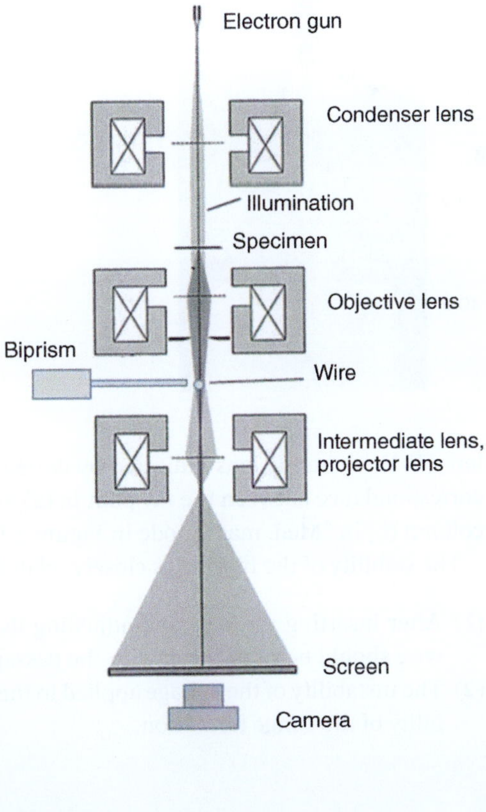

Figure 5.7 Block diagram showing main components of transmission electron microscope and standard biprism position.

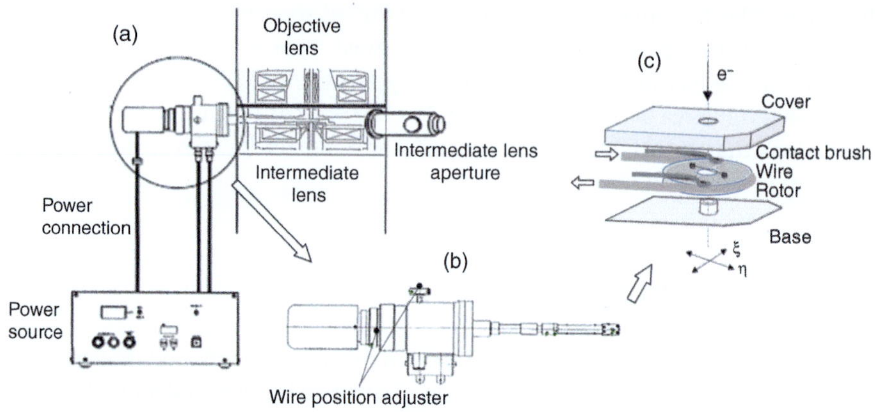

Figure 5.8 Biprism wire holder and power connection. (a) Outline of biprism wire holder and power connection to electron microscope column. (b) Biprism wire holder indicating screws used to adjust wire position relative to optical axis. (c) Mechanism for adjusting wire direction relative to specimen by rotating biprism wire.

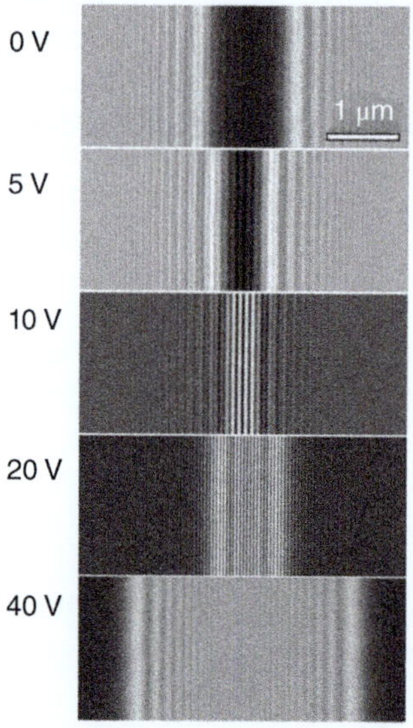

Figure 5.9 Wire voltage vs. fringes with accelerating voltage of 300 kV. According to Eq. (5.39), fringe spacing is inversely proportional to wire voltage, and fringe field width is proportional to wire voltage.

lens (IL1) where the lens indicated by dotted lines is set off. Figure 5.11 shows the correspondence between the ray path in (a) and the cross section of the microscope column (b) in "Med. mag" mode in Figure 5.10.

The stability of the biprism is closely related to several factors as follows:

(1) After inserting the wire and adjusting its position, the mechanical drift of the wire should be suppressed with the passage of time.
(2) The instability of the voltage applied to the biprism is directly linked to the instability of the fringe formation.

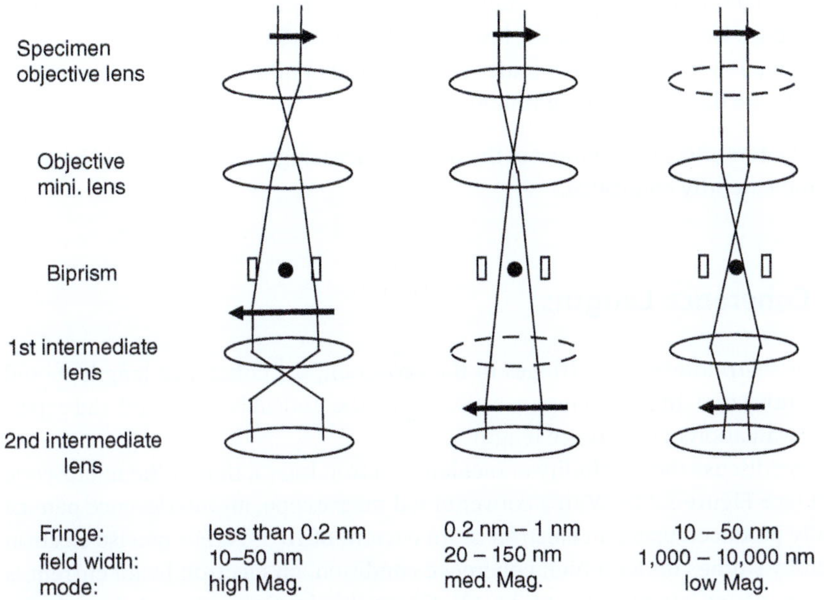

Figure 5.10 Ray paths of electron beam passing through specimen, biprism wire, and hologram positions for three magnification ranges.

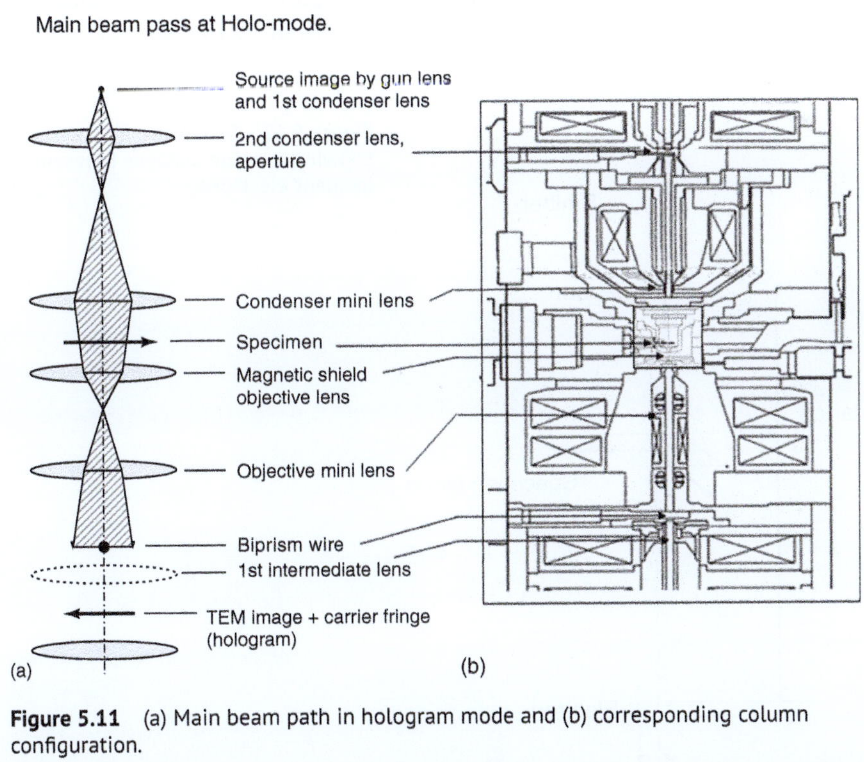

Figure 5.11 (a) Main beam path in hologram mode and (b) corresponding column configuration.

(3) The instability of the acceleration voltage corresponding to the limitation of coherence length results in the low visibility of interference fringes.
(4) Illumination for a long time under poor vacuum condition results in the charging of wire due to the contamination.

These factors affect the contrast of the interference fringes, so they should be monitored and carefully controlled.

5.3 Coherence Lengths

To obtain sharp interference fringes in the hologram, the coherence length should be efficiently large. In this section, we discuss the theoretical background and experimental evaluation of coherence length.

First, we discuss the possibility of incident electron interaction in the microscope column (see Figure 5.12). With a conventional microscope, an interference pattern is usually produced using an electron beam current of 10^{-10} A. For precise electron holography studies under a high coherence condition, an electron beam current is reduced to about 2.0×10^{-11} A at 200 kV. Given this low current and the velocity of 200-keV electrons (2.08×10^8 m s^{-1}), the average separation between electrons is about 3 m. Since the column is less than 1.7 m long, there is on average only one electron in the column at a time. Thus, as shown in Figure 5.12, it is considered that interactions among the incident electrons do not contribute to the electron interference effect.

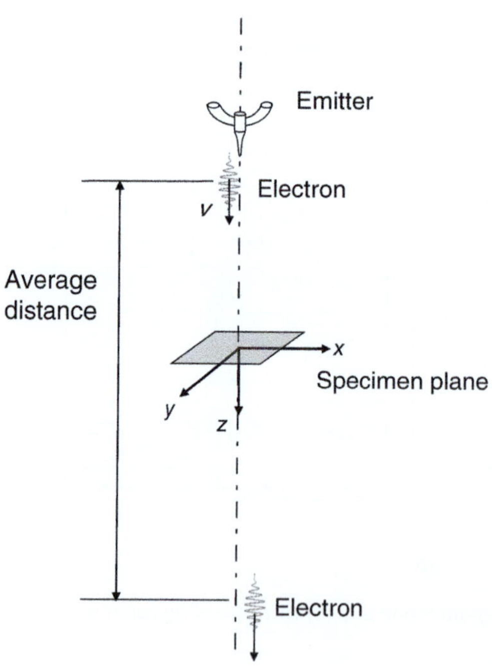

Figure 5.12 Schematic illustration showing average distance between incident electrons.

Experimentally, the effective coherence length of an electron plane wave has been treated in a way similar to that used in light optics [3]. An ideal wave source is considered to be a monochromatic point source, but in reality, the wave source is extended and polychromatic. The effects of the finite size of the source can be discussed in terms of spatial or lateral coherence length L_{SP} while the effects of different energies can be discussed in terms of temporal or longitudinal coherence length L_{TE}. These effective coherence lengths of the plane wave have been given as follows [4]:

$$L_{SP} \simeq \lambda/(2\alpha_i) \quad (5.4)$$

$$L_{TE} \simeq (2E/\Delta E)\cdot\lambda \quad (5.5)$$

where L_{SP} and L_{TE} correspond to the width and length of the wave packet.

In Figure 5.13, α_i is the illumination angle in Eq. (5.4), which can be estimated by measuring the radius of an electron diffraction disk observed under the same illumination condition. Using the relation among the illumination angle (α_i), the brightness of the emitter (β) and its current density (J), i.e. $\beta = J/(\pi\alpha_i^2)$, the spatial coherence length is given by β and J. Figure 5.14 shows the relation between L_{SP} and J at 200 kV with β as a parameter.

In Eq. (5.5), E and ΔE are the kinetic energy and energy spread of the incident electrons, respectively; ΔE can be obtained directly from the energy-loss spectrum. In a 200-kV electron microscope equipped with a thermal FEG, $\lambda = 0.0025$ nm and $\Delta E \simeq 0.8$ eV; in conventional electron diffraction, $\alpha_i \simeq 10^{-6}$ rad. Using these values, we obtain $L_{SP} \simeq 1.3$ μm and $L_{TE} \simeq 1.3$ μm.

Coherence width L_{SP} is considered to correspond to the uncertainty of the position (Δr) of an incident electron while the uncertainty of the momentum (Δp) is given as

$$\Delta p = \alpha_i \cdot p = \alpha_i\, h/\lambda \quad (5.6)$$

Thus, Eq. (5.4) corresponds to the Heisenberg uncertainty principle

$$\Delta r \cdot \Delta p = h/2\,(\geq h/4\pi) \quad (5.7)$$

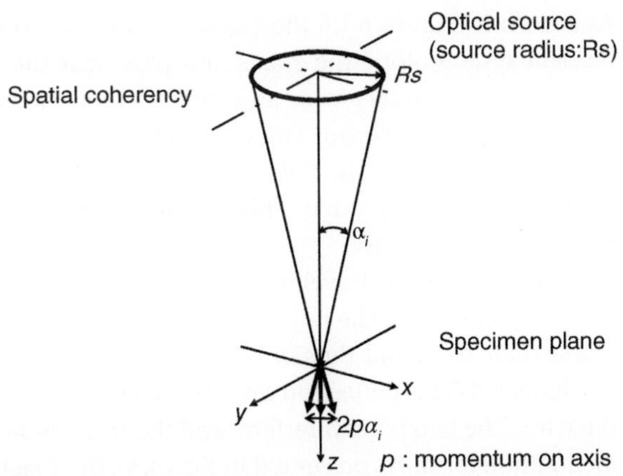

Figure 5.13 Illumination angle and vertical component of electron momentum relative to optical axis.

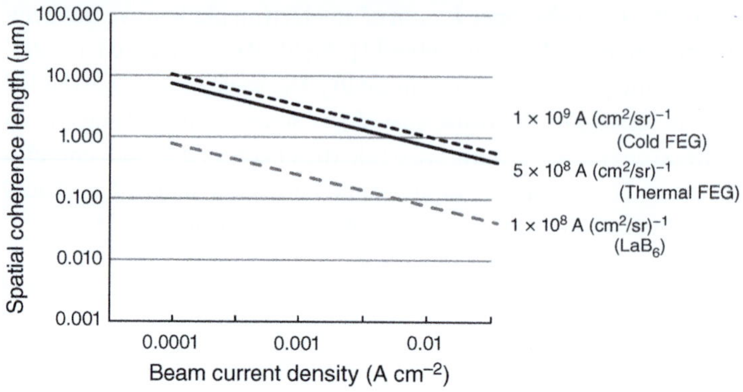

Figure 5.14 Calculated beam current density vs. spatial coherence length at 200 kV with source brightness as parameter.

5.4 Formation of Interference Fringes

The formation of interference fringes is the most important process in recording electromagnetic field information. In Section 5.4.1, the simple interpretation of interference effect between two plane waves assumed to originate from two virtual sources is explained. This explanation enables the basic principles of the fringe formation to be simply understood. In Section 5.4.2, a more sophisticated explanation based on quantum mechanics with Green's function is presented. The treatment presented by Komrska [5] enables the features of interference fringes to be accurately captured and explains the fundamental principles of detailed interference fringe formation. This treatment is used in the simulation of interference fringes (Section 5.5) and is theoretically discussed in Chapter 12.

5.4.1 Geometrical-Path Interpretation with Two Virtual Sources

As shown in Figure 5.15, the electrons emitted from the electron source enter the biprism system along the z-axis and pass near the wire. The electron wave starts from the source and spreads as a spherical wave for which the radius reaches about 100 mm near the biprism. The spread in the direction perpendicular to the electron traveling direction in the field of view is at most several μm, so the curvature of the wavefront can be neglected. This means that the electron wave can be approximately treated as a plane wave.

The electron wave, traveling along the optical axis, is deflected by the electric field of the biprism wire. The ratio of the biprism wire voltage to the acceleration voltage is less than 10^{-3}, and thus only the electrons passing near the surface of the wire are deflected. This deflection divides the beam into two parts, to the right and left of the wire. The two parts interfere, and the interference fringes responsible for phase modulation form. As presented in Eq. (4.5), the spacing of the interference fringes d

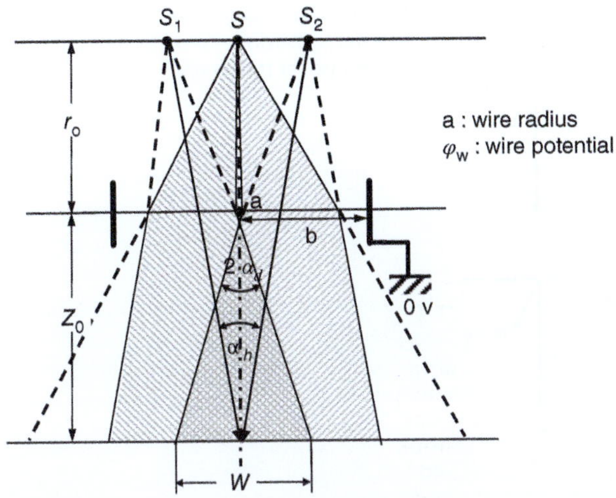

Figure 5.15 Schematic showing how main beam is split into two parts by biprism wire (diameter 2a). Two parts overlap at an angle as two plane waves. Interference fringes are produced in overlap area, which has width *W*.

is given by:

$$d = \frac{\lambda}{\alpha_h} \tag{5.8}$$

where α_h is a superimposed angle between the object wave and the reference wave as illustrated in Figure 5.15. In the next section, d is given taking into account the geometrical condition with the electric field of wire. According to Eq. (5.39), α_h is given as

$$\alpha_h = \frac{\lambda}{d} = -\frac{r_0}{(r_0 + z_0)} \frac{\pi e \varphi_w}{E} \frac{1}{\ln(a/b)} \tag{5.9}$$

(Note $\ln(a/b) < 0$)
where e is the elementary electric charge, E is the kinetic energy of the incident electron, and φ_w is the wire potential. According to the illustration in Figure 5.15, the relation between the superposed angle α_h and the deflection angle α_d is given as:

$$\alpha_h = \frac{2r_0\alpha_d}{r_0 + z_0} \tag{5.10}$$

The width of the interference fringes W is also given with α_d as shown in Figure 5.15

$$W \sim 2z_0\alpha_d - \frac{2a(r_0 + z_0)}{r_0} \tag{5.11}$$

5.4.2 Wave-Optical Treatment

In the interpretation above, the direction of electron travel was simply assumed to be the direction of the wave vector of the plane wave, and the wave surface was assumed to be divided into two plane waves by the biprism wire. The two plane waves were

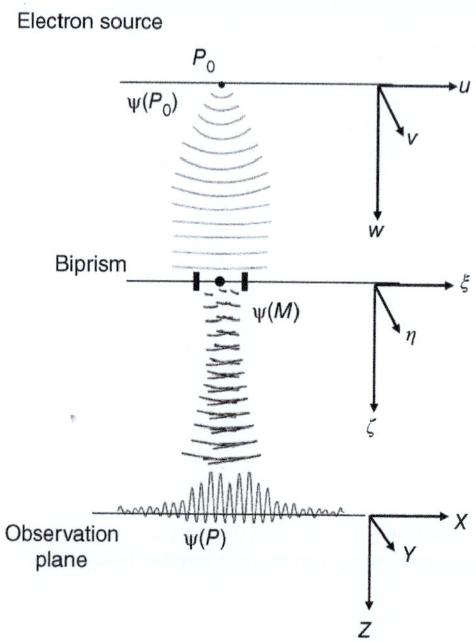

Figure 5.16 Illustration showing interference fringe formation based on wave-optical treatment.

assumed to be tilted by the same deflection angle along the electron path by the wire electrical field.

In this section, the incident electron is represented as a wave function on the basis of quantum mechanics [5]. As shown in Figure 5.16, in this wave-optical treatment, the electron source (P_0) is located far enough from the biprism wire (several cm). As shown in Figure 5.17, the biprism wire field drops quickly to a few micrometers as the distance increases. This means that the spherical wave $\psi(P_0)$ emitted from P_0 can be treated as a plane wave at the biprism position as it was in the case above. In the case here, there is a single-electron source, not two virtual electron sources. The plane wave is represented as wave function $\psi(M)$ affected by the electric field of the biprism and is represented as wave function $\psi(P)$ at the observation point. Therefore, our objective is to obtain wave function $\psi(M)$ at the biprism position from wave function $\psi(P_0)$ at the electron source and to express the wave function of the observation point using $\psi(M)$. The coordinate system at each position used in the discussion here is shown in Figure 5.18.

5.4.2.1 Wave Function at Wire Plane

Let $\psi_U(M)$ be the solution of Schrödinger's equation of the wave function influenced by the electric field of the wire. Using $f(M)$ as a transmission function, we can represent $\psi(M)$ as

$$\psi(M) = \psi_U(M) \cdot f(M) \tag{5.12}$$

where $f(M)$ represents only the amplitude and $\psi_U(M)$ is given with the path (S) as a function of $\vec{r}$ and E:

$$\psi_U(\vec{r}) = A(\vec{r}) \exp[ikS(\vec{r}, E)] \tag{5.13}$$

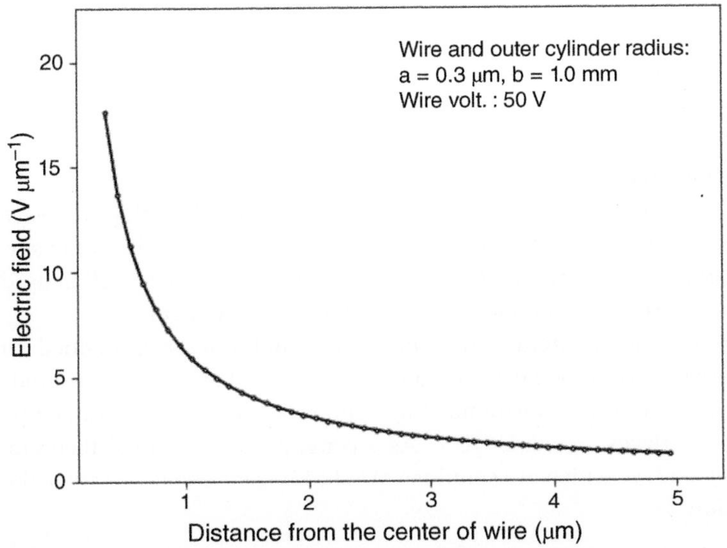

Figure 5.17 Electric potential as function of distance from center of biprism wire.

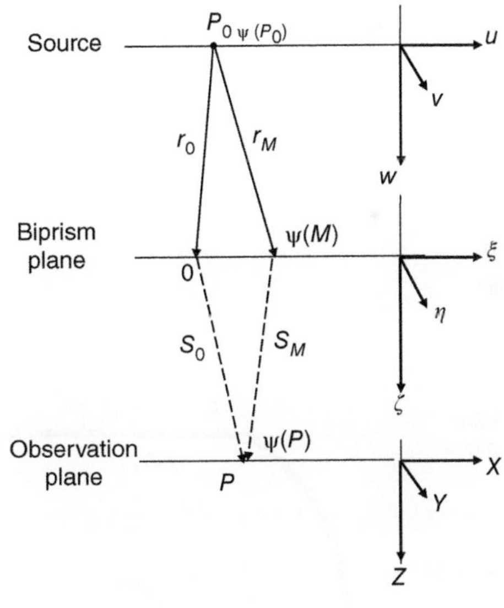

Figure 5.18 Illustration showing ray paths from electron source to observation plane.

If we assume that $\psi_U(\vec{r})$ satisfies Schrödinger's equation, $S(\vec{r},E)$ can be expressed as

$$S(\vec{r},E) = l - \frac{1}{2E}\int_l U(\vec{r})dl \tag{5.14}$$

where "l" is the integral along the classical path. We set

$$A(\vec{r}) = |\psi_0(\vec{r})| \tag{5.15}$$

and then obtain

$$\psi_U(\vec{r}) = |\psi_0(\vec{r})| \exp\left\{ i\vec{k} \cdot \vec{r} \left[1 - \frac{1}{2E} \int_0^1 U(P_0 + \vec{r}\,t)dt \right] \right\} \tag{5.16}$$

5.4.2.2 Green's Integral Theorem

Wave function $\psi(P)$ at the observation point is obtained by calculating the wave motion generated at the observation point on the basis of the waves from the surrounding wave sources. Here, we use Green's integral theorem, which is often used for such purposes, to solve this problem as a boundary value problem.

First, we discuss the theorem using the spaces and interfaces defined in Figure 5.19: biprism surface plane S_1 contains the wire, and sphere S' surrounds observation point P. Point P is surrounded by volume V, which is surrounded by these connection surfaces: $S = S_1 U S_2 U S'$. As a consequence of Gauss' theorem, wave functions ψ and ψ_1 which are solutions of Schrödinger's equation satisfy the following equation,

$$\iiint_V (\psi\nabla^2\psi_1 - \psi_1\nabla^2\psi)dV = \iint_s (\psi\nabla\psi_1 - \psi_1\nabla\psi)\vec{n}dS \tag{5.17}$$

where $\vec{n}$ is a unit vector perpendicular to the surface. Since ψ and ψ_1 are solutions of Schrödinger's equation, the term in the volume integral is zero. According to

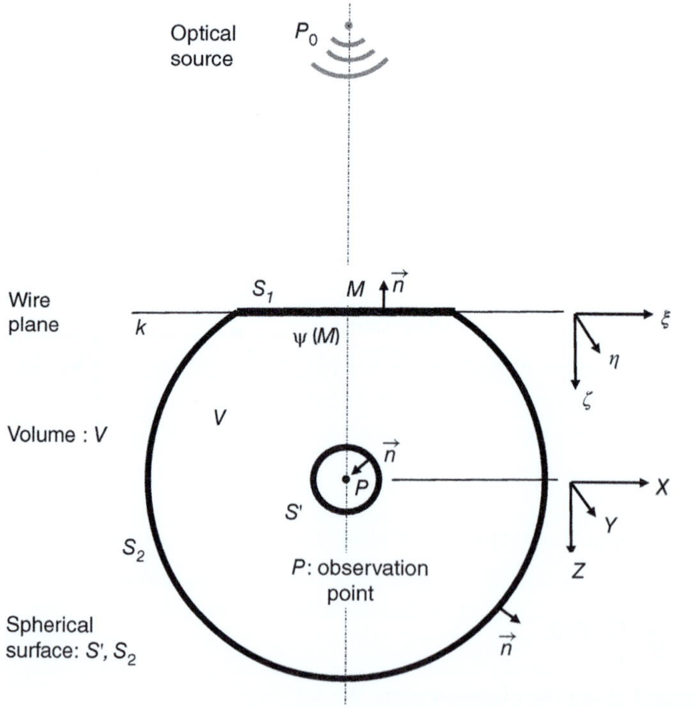

Figure 5.19 Spaces and interfaces defined for discussion of Green's integral theorem.

Sommerfeld, we apply the Green's function G to ψ_1 [6]. Thus, we get the relation

$$\iint_{S_1 \cup S_2} (\psi \nabla G - G \nabla \psi)\vec{n} dS + \iint_{S'} (\psi \nabla G - G \nabla \psi)\vec{n} dS = 0 \tag{5.18}$$

Furthermore, the integral of surface S' including observation point P is the extremum of the surface area surrounding P when S' approaches 0 while moving away from P. Thus, in the limit of the integral over sphere S' in Eq. (5.18), we get

$$\lim_{s \to 0} \iint_{S'} (\psi \nabla G - G \nabla \psi)\vec{n} dS = \psi(P) \int_0^{2\pi} \int_0^{\pi} \sin \vartheta d\vartheta d\chi = 4\pi \psi(P) \tag{5.19}$$

Substituting this into (5.18), we get

$$\begin{aligned}\psi(P) &= -\frac{1}{4\pi} \iint_{S_1 \cup S_2} (\psi \nabla G - G \nabla \psi)\vec{n} dS \\ &= -\frac{1}{4\pi} \iint_{S_1} \psi \nabla G \vec{n} dS - \frac{1}{4\pi} \iint_{S_2} (\psi \nabla G - G \nabla \psi)\vec{n} dS\end{aligned} \tag{5.20}$$

where we used $G(P,M) = 0$ at points $Q \neq P$ of the half-space (κ,P). When the radius (s) of S_2 approaches infinity, the integration of S_2 is almost 0, i.e.

$$\lim_{x \to \infty} \frac{1}{4\pi} \iint_{S_2} (\psi \nabla G - G \nabla \psi)\vec{n} dS = 0 \tag{5.21}$$

and only the integral for S_1 remains,

$$\psi(P) = -\frac{1}{4\pi} \iint_{S_1} \psi(M) \nabla G(P, M)\vec{n} dS \tag{5.22}$$

Thus, wave function $\psi(P)$ is obtained with $\psi(M)$ which is affected by the electric field on wire surface, with Green's function $G(P, M)$ which represents the propagation of the wave from point M to observation point P. To evaluate the wave function $\psi(P)$ at the observation plane, we further clarify Green's function as shown below.

5.4.2.3 Explicit Form of Green's Function

Green's function $G(P, M)$ in Eq. (5.22) shows the propagation of the wave from the wire plane to the observation plane. By means of the method of images [6], Green's function $G(P, Q)$ for points M and P shown in Figure 5.20 is given by

$$\begin{aligned}G(P, Q) = &\frac{1}{s_Q} \exp\left\{ iks_Q \left[1 - \frac{1}{2E} \int_0^1 U_g(P + \vec{s}_Q t) dt \right] \right\} \\ &- \frac{1}{s_{Q'}} \exp\left\{ iks_{Q'} \left[1 - \frac{1}{2E} \int_0^1 U_g(P' + \vec{s}_{Q'} t) dt \right] \right\}\end{aligned} \tag{5.23}$$

where the function $U_g(\xi, \eta, \zeta)$ defined by

$$U_g(\xi, \eta, \zeta) = U(\xi, \eta, \zeta) \qquad \text{if } \xi \geqq 0,$$

$$U_g(\xi, \eta, \zeta) = U(\xi, \eta, -\zeta) \qquad \text{if } \zeta \leqq 0.$$

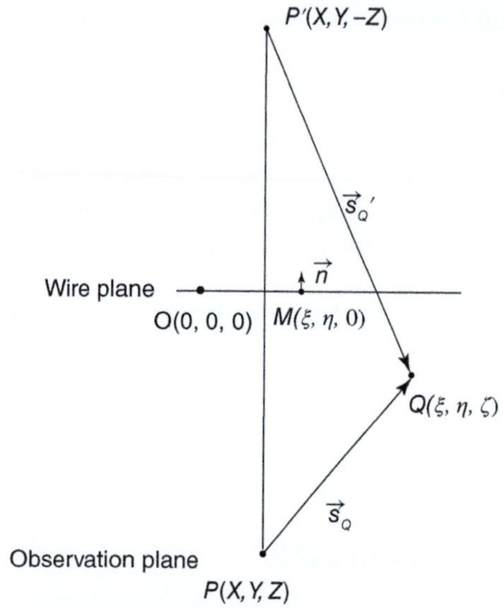

Figure 5.20 Geometrical configuration for discussion of Green's half-space function.

The first term in Eq. (5.23) represents the spherical wave from point Q to point P (observation point), and the second term represents the spherical wave from point P' (point symmetric to observation point across wire plane) to point Q. As shown in Figure 5.20, P is the observation point, P' is the symmetry point of P with respect to the wire plane, and Q is the space on the side of the two half-spaces divided by the wire plane that includes the observation point. Using the conditions $1/(k \cdot s_M) \ll 1$, $U/2E \ll 1$, we obtain

$$\nabla G \cdot \vec{n} = 2ik \cdot \exp\left\{ iks_M \left[1 - \frac{1}{2E}\int_0^1 U(P + \vec{s}_M t)dt\right]\right\} \frac{\vec{s}_M \cdot \vec{n}}{s_M{}^2} \tag{5.24}$$

where E is electron energy, U is the potential at the electron path, k is the wave number, s_M is the path from the biprism to the observation plane, and t is a parameter.

5.4.2.4 Intensity Distribution of Interference Fringes

The wave function on observation plane $\psi(P)$ as given in Eq. (5.22) can now be given by inserting $\psi(M)$ from Eq. (5.12) and ∇G from Eq. (5.24) as

$$\psi(P) = D \iint_S f(M) \exp\{i[\phi(M) + \phi_U(M)]\} dS_M \tag{5.25}$$

where the amplitude and phase are as follows. The wave paths from the electron source to the observation plane are shown in Figure 5.18:

$$D = -\frac{ikA_0}{2\pi r_0 s_0} \exp[ik(r_0 + s_0)] \cdot \exp\left[-\frac{ik}{2E}\int_0^1 [r_0 U(P_0 + r_0 t) + s_0 U(P + s_0 t)]dt\right]$$
$$(|\psi_0(r_M)| = A_0/r_M) \tag{5.26}$$

$$\phi(M) = k[r_M + s_M - (r_0 + s_0)] \tag{5.27}$$

$$\phi_U(M) = -\frac{k}{2E}\int_0^1 [r_M U(P_0 + \vec{r}_M t) + s_M U(P + \vec{s}_M t)$$
$$-r_0 U(P_0 + \vec{r}_0 t) - s_0 U(P + \vec{s}_0 t)]dt \tag{5.28}$$

These equations above are used to calculate interference fringe intensity. The A_0 in Eq. (5.26) corresponds to the amplitude of the source.

We can now express $\psi(P)$ as

$$\psi(P) = D_2 \left[\int_{-\xi_{max}}^{-a} + \int_a^{\xi_{max}}\right] \exp\{i[\phi(\xi) + \phi_U(\xi)]\}d\xi \tag{5.29}$$

According to cylindrical capacitor potential shown as Figure 5.21, where D_2 and the exponential term are given as

$$D_2 = \frac{\exp(-i\pi/4)}{\pi}\frac{A_0}{(r_0 z_0)^{1/2}} \exp\left[ik\left(R - b\frac{|e|\varphi_W}{\text{In}(a/b)}\frac{1}{E}\right)\right] \tag{5.30}$$

$$\phi(\xi) + \phi_U(\xi) = k\left\{\frac{1}{2}\left(\frac{1}{r_0} + \frac{1}{z_0}\right)\left(\xi - \frac{xr_0}{r_0 + z_0}\right)^2 + \frac{|e|\varphi_W}{\text{In}(a/b)}\frac{1}{2E}\right.$$
$$\left.\cdot\,\xi\left[\left(\frac{\xi}{r_0} + \frac{\xi - x}{z_0}\right)\text{In}\frac{|\xi|}{b} + \pi\cdot\text{sgn}\xi - \frac{\xi}{b}\left(1 + \frac{1}{12}\frac{\xi^2}{b^2}\right)\right]\right\} \tag{5.31}$$

Using $\psi(P)$ defined in Eq. (5.29), we can finally express the relative intensity distribution of the interference fringes at the observation plane as

$$I(X) = \frac{\psi^*\psi}{\psi_0{}^*\psi_0} = \frac{k}{2\pi}\left(\frac{1}{r_0} + \frac{1}{z_0}\right)\left|\left[\int_{-\xi_{max}}^{-a} + \int_a^{\xi_{max}}\right]\exp\{i[\phi(\xi) + \phi_U(\xi)]\}d\xi\right|^2 \tag{5.32}$$

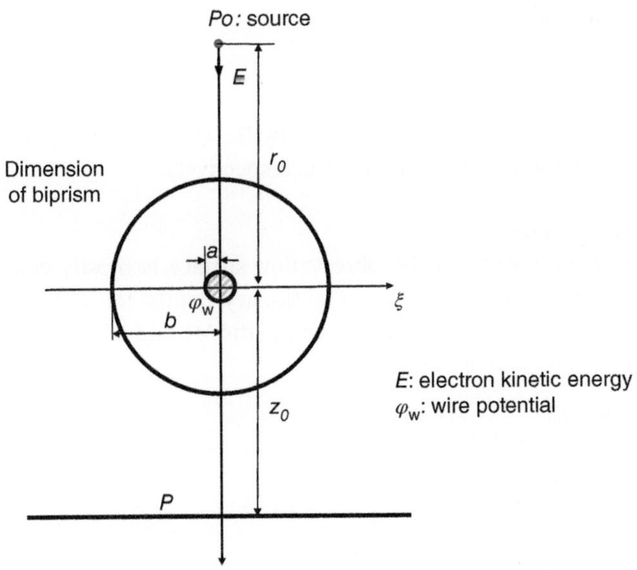

Figure 5.21 Geometry for calculation of interference fringe intensity.

The formulations derived in this section are used below to characterize the features of the interference fringe distribution such as the stationary points, interference region, and interference fringe spacing. Furthermore, they are used to simulate the detailed interference fringe distribution, which is compared with the observed fringe distribution in Section 5.5.

5.4.2.5 Stationary Points and Interference Region

If we differentiate Eq. (5.31) by coordinate ξ of the wire plane and set it to 0,

$$\frac{d}{d\xi}\{\phi(\xi) + \phi_U(\xi)\} = 0 \tag{5.33}$$

From Eq. (5.33), we can obtain points ξ_1 and ξ_2 on the wire plane:

$$\xi_1 = \frac{r_0 x}{r_0 + z_0} - \pi \frac{r_0 z_0}{r_0 + z_0} \frac{e\varphi_W}{\text{In}(a/b)} \frac{1}{2E} \tag{5.34}$$

$$\xi_2 = \frac{r_0 x}{r_0 + z_0} + \pi \frac{r_0 z_0}{r_0 + z_0} \frac{e\varphi_W}{\text{In}(a/b)} \frac{1}{2E} \tag{5.35}$$

$(\ln(a/b) < 0)$

This means that there are two stable paths with a small phase change. The contributions of these two paths mainly result in the formation of interference fringes. This point is discussed further in the last part of Chapter 12 (see Figure 12.7). When ξ_1 or ξ_2 reaches the end of the wire, the fringe disappears because the wave of the stable phase forming the fringe is cut by the wire. The points on the observation surface at this time are X_A and X_B:

$$X_{A,B} = \pm\left[\pi z_0 \frac{e\varphi_W}{\text{In}(a/b)} \frac{1}{2E} + \frac{a}{r_0}(r_0 + z_0)\right] \tag{5.36}$$

The fringe field width is thus given by

$$W = X_B - X_A = -2\left[\pi z_0 \frac{e\varphi_W}{\text{In}(a/b)} \frac{1}{2E} + \frac{a}{r_0}(r_0 + z_0)\right] \tag{5.37}$$

Thus, width W of the interference region is obtained by Eq. (5.37) from X_A and X_B, with $\xi_1 = a$ and $\xi_2 = -a$ at the two ends of the wire, respectively.

5.4.2.6 Spacing of Interference Fringes

Since the wave function of the points on the observation surface is mostly contributed by the waves ($\Psi(M)$ in Figure 5.18) from stationary points *M1* and *M2*, it reaches maximum when the phase difference between paths *M1* and *M2* is 0, $\pm\pi$, $\pm 2\pi, \ldots$. Given this condition, we can define

$$X_n = -n\frac{r_0 + z_0}{r_0} \cdot \text{In}(a/b) \cdot \frac{\lambda E}{2\pi e\varphi_w}(\ln(a/b) < 0) \tag{5.38}$$

Thus, the fringe interval d is given by

$(\ln(a/b) < 0)$

$$d = X_{n+2} - X_n = -\frac{(r_0 + z_0)}{r_0} \cdot \text{In}(a/b) \cdot \frac{\lambda E}{\pi e\varphi_w} \tag{5.39}$$

The detailed interference fringe distribution with Fresnel fringes can be obtained with the formulations derived in this section, as shown in the following section.

5.5 Simulation of Interference Fringes

Here, using the formulations derived in Section 5.4.2, we compare the measured data with the simulated results. The interference fringes observed using a CCD camera with a 300-kV transmission electron microscope are shown in Figure 5.22a. The intensity profile of the fringes in band region A–B in (a) is shown in (b). The normalized intensity profile of the interference fringes obtained with Eq. (5.32) is shown in (d). The parameters and the values used in the simulation are shown in Table 5.1. The fringe spacing and intensity profile of the observed interference fringes including the Fresnel fringes at both ends of the intensity profile (Figure 5.22b) are reproduced in the simulated profile. As described in Section 5.4.1, in the case of geometrical-path treatment, the intensity modulation with Fresnel fringes cannot be obtained.

To take into account the contribution of the illumination angle and the energy spread of the incident electrons corresponding to the effective coherence lengths discussed in Section 5.3, the simulated fringes shown in Figure 5.22d were broadened by superimposing slightly shifted profiles. The resulting profile in Figure 5.22c shows good agreement with the observed one in Figure 5.22b.

The process of interference fringe formation with an increase in incident electrons, as shown in Figure 5.23, cannot be explained with a simple plane wave incidence. This is discussed in detail using a simulation model in the Chapter 12.

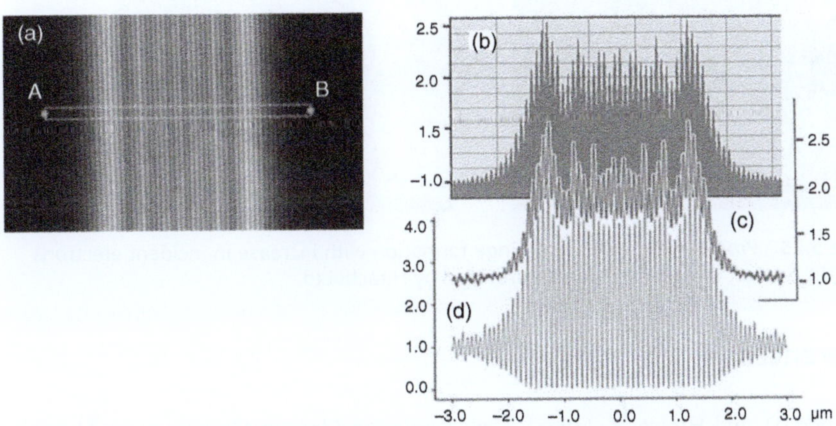

Figure 5.22 Comparison of profiles of observed and simulated interference fringes. (a) Intensity distribution of interference fringes observed at wire potential 35 V and with accelerating voltages of 300 kV, respectively. (b) Intensity profile of interference fringes averaged in vertical direction for region A–B in (a). (c and d) Intensity profile of simulated interference fringes. (c) Intensity profile of interference fringes obtained with intensity profile by taking into account contribution of illumination angle and energy spread of incident electrons. (d) Normalized intensity profile of interference fringes simulated with parameters in Table 5.1.

Table 5.1 Parameters used for hologram simulation.

Parameter	Symbol	Value used for simulation
Energy of electrons	E	300 keV
Wave number	k	1.95×10^7 mm^{-1}
Source to wire	r_0	200 mm
Wire to observation plane	z_0	71 mm
Wire diameter	$2a$	5×10^{-4} mm
Outer cylinder diameter	$2b$	1 mm
Wire potential	φ_w	35 V

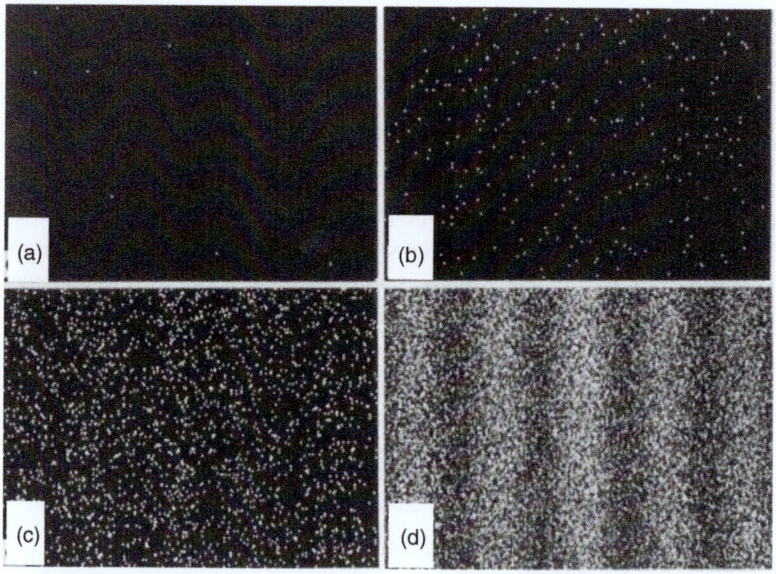

Figure 5.23 Process of interference fringe formation with increase in incident electrons ((a)→(d)). Source: Images were kindly provided by Hitachi Ltd.

References

1 Shindo, D. and Hiraga, K. (1998). *High Resolution Electron Microscopy for Materials Science*. Tokyo: Springer.

2 Moony, P.E., Fan, G.Y., Meyer, C.E. et al. (1990). Slow-Scan CCD Camera for Transmission Electron Microscopy. In: *Proceedings, 12th International Congress for Electron Microscopy*, Seattle, San Francisco Press, San Francisco, 164–165.

3 Born, M. and Wolf, E. (1964). *Principles of Optics*, 2e. Oxford: Pergamon Press.

4 Heidenreich, R.D. (1964). *Fundamentals of Transmission Electron Microscopy*. New York: Interscience.

5 Komrska, J. (1971). *Scalar diffraction theory in electron optics. Adv. Electron. Electron Phys.* 30: 139.

6 Sommerfeld, A. (1964). *"Optics", Section 34*. New York: Academic Press.

6

Related Techniques and Specialized Instrumentation

6.1 Split-Illumination Electron Holography

In conventional electron holography, if a strong stray field exists near the specimen, the reference wave is severely modulated, meaning that the condition given by Eq. (4.5) for observing a simple interference phenomenon is not satisfied. To solve this problem, split-illumination electron holography was developed. As shown in Figure 6.1a, a biprism was added in the illumination system to split the coherent electron wave into two coherent waves [2]. One of the split waves is used as the object wave illuminating an observation area far from the specimen edge, and the other is used as the reference wave passing through the vacuum outside the specimen.

These waves are interfered by using two biprisms in the imaging system to form a hologram. A high-contrast hologram can be obtained in an area far from the specimen edge without reducing the illuminating electron intensity accordingly. Placing the reference area away from the specimen can also be done to reduce modulation of the reference wave due to stray fields from the specimen. The separation distance between the two waves is controlled by adjusting the voltage of the biprism in the illumination system.

Fringe spacing s and hologram width W are independently controlled by adjusting the voltages applied to the upper and lower biprisms in the imaging system, as in double-biprism electron interferometry [3]. The filament electrode of the upper biprism is placed at the image plane of the objective lens while that of the lower biprism is placed between the crossover point below the magnifying lenses and the hologram plane (see Figure 6.1a).

For precise phase measurement, further improvements were achieved by using double-biprism split-illumination electron holography without Fresnel fringes, as shown in Figure 6.1b [4]. The upper and lower condenser biprisms are placed between the second and third condenser lenses. The third lens is used to focus the shadow of the upper condenser biprism filament onto the specimen plane. As a result, the Fresnel fringes produced by the upper condenser biprism are not superimposed on the hologram. The lower biprism is placed in the shadow formed between the paths of the two coherent electron waves separated by the upper condenser biprism. The improved system is capable of imaging by using only a single biprism in the imaging system.

Material Characterization using Electron Holography, First Edition. Daisuke Shindo and Takeshi Tomita.

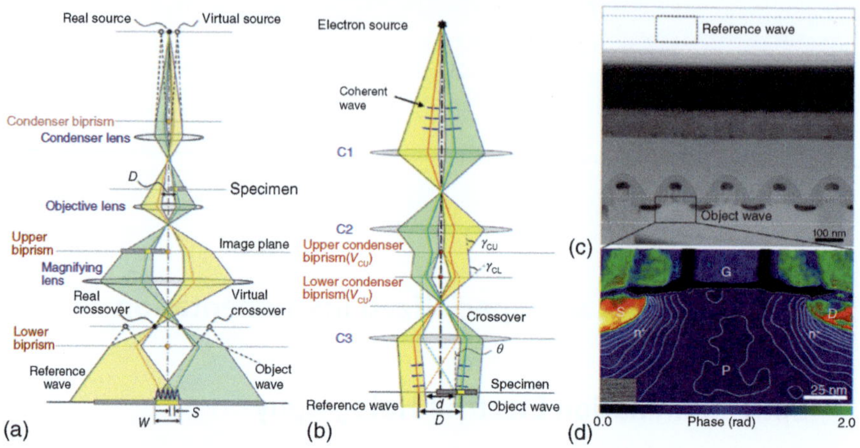

Figure 6.1 (a) Ray path for (single-biprism) split-illumination electron holography. (b) Optics of illumination system for (double-biprism) split-illumination electron holography. (c) TEM image of nMOSFET with 50-nm gate length. (d) Phase image of FET area indicated by rectangle in (c). Inset at bottom left of panel shows part of hologram, in which white contour lines have a phase difference of 0.07 rad, corresponding to an electrostatic potential difference of 0.1 V. Source: Shindo and Akase [1], with permission from Elsevier.

This improved system was used to observe the electric potentials in a semiconductor device [4]. Thin specimens were prepared using the conventional method from a Cu-processed n-type metal–oxide–semiconductor field-effect transistor (nMOSFET) chip prepared with a focused ion beam (FIB) instrument. The gate length was 50 nm, and the carrier concentration under the source and drain regions was designed to be 6×10^{20} cm^{-3} by doping As into a p-type silicon substrate. The amount of doping in the p-doped substrate area was 1×10^{17} cm^{-3}.

As shown in Figure 6.1c, the transistor area is far from the specimen edge. A double-biprism split-illumination system was used to set the area of the reference wave in the vacuum region indicated by the rectangle. The phase shift was obtained using Eq. (4.2), with $\vec{A}(x, y) = 0$. Since the specimen thickness is constant, phase shift $\phi(x, y)$ is simply given by $\sigma\varphi(x, y)$t, where t is the specimen thickness.

Figure 6.1d shows the relative phase distributions in the transistor area with a spatial resolution of 3 nm (interference fringe spacing of 1 nm) and phase noise (standard deviation in Si substrate) of ± 0.03 rad for the very long distance of 640 nm between the object and reference waves. The inset at the lower left shows a part of the hologram indicating the sharp interference fringes observed with the improved illumination system. The electric potential differences were calculated from the measured specimen thickness of 95 ± 5 nm, which was measured by convergent beam electron diffraction (CBED) [5–9]. An electric potential difference of 0.1 V corresponds to the contour lines in the phase image, which have a phase difference of 0.07 rad. The measured electric potentials in the source and drain regions were 1.1 V higher than that in the substrate. This result agrees well with the electrostatic potential distributions in this nMOSFET design.

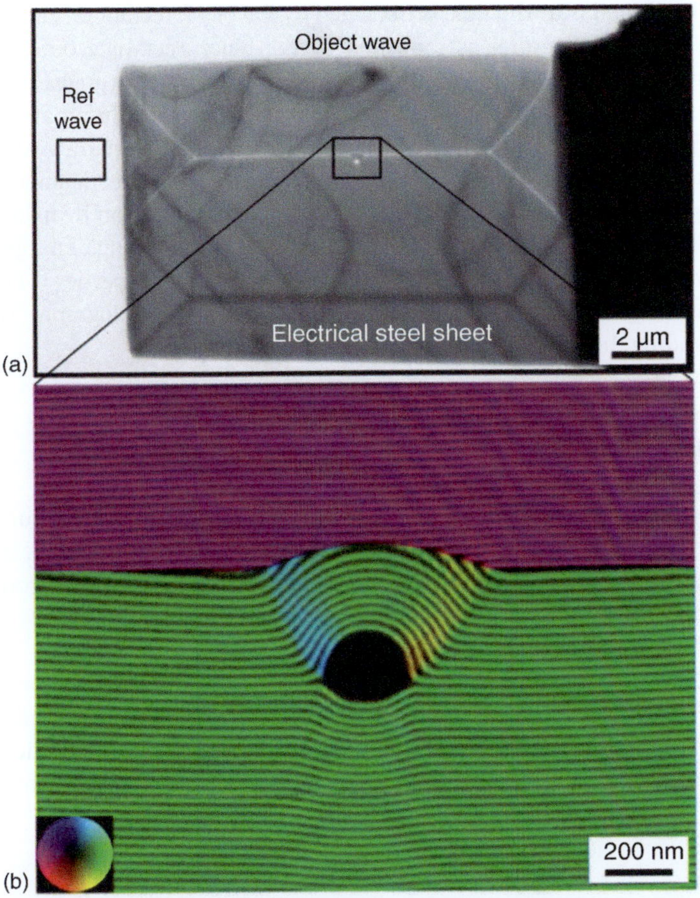

Figure 6.2 Observation of magnetic flux in electrical steel sheet with sub-micrometer hole. (a) Over-focused Lorentz microscopy image. (b) Magnetic flux lines (green and crimson) showing magnetic domain wall pinned around hole. Source: Shindo and Akase [1], with permission from Elsevier.

Figure 6.2 shows an application of split-illumination electron holography to magnetic field analysis of a thin, doubly oriented electrical steel sheet [2], which is a key component of power generators, motors, and transformers. The magnetic properties of this material have been studied using Lorentz microscopy to observe pinning phenomena [10] and the dynamic behavior of the domain walls at the precipitates and around a hole created as an artificial nonmagnetic area [11]. The use of split-illumination electron holography enables quantitative measurement of the magnetic flux distribution around defects far from the specimen edge. A thin specimen with an ellipsoidal hole was prepared using an FIB system. The crystallographic orientations of the [001] and [110] directions were approximately parallel to the long and short sides, respectively, of the rectangular specimen.

Figure 6.2a shows a Lorentz microscopy image observed in an over-focus condition with the straight black and white lines indicating the domain walls. The area of the domain walls pinned near the hole is indicated by the black rectangle within the image, while the reference area is indicated by the black rectangle outside the specimen. Figure 6.2b shows the quantitative magnetic flux lines (green and crimson) around the hole (black area), which were obtained by split-illumination electron holography. The detailed magnetic flux behavior at the domain walls is clearly evident [2]. These findings show that we can estimate the magnetic energy of a system containing defects from the magnetic flux distribution by using micromagnetic simulation. The magnetic energies, such as the exchange energy, anisotropy energy and demagnetization energy affect the magnetic properties [11]. Thus, the detailed magnetic flux distribution observed by electron holography is useful in understanding the magnetic properties.

6.2 Dark-Field Electron Holographic Interferometry

Dark-field electron holographic (DFEH) interferometry was developed for measuring the strain field for a relatively wide field of view and a thick specimen [12]. It combines the advantage of moiré imaging techniques and the flexibility of electron holography. Figure 6.3 shows the geometrical configuration used for this interferometry technique compared with that of conventional electron holography.

As shown in Figure 6.3b, in DFEH interferometry, a coherent electron beam illuminates the specimen in a diffraction condition for a certain set of lattice planes. The specimen is composed of a zone of unstrained crystal with known lattice parameters, adjacent to a zone of strained crystal, in a similar orientation and

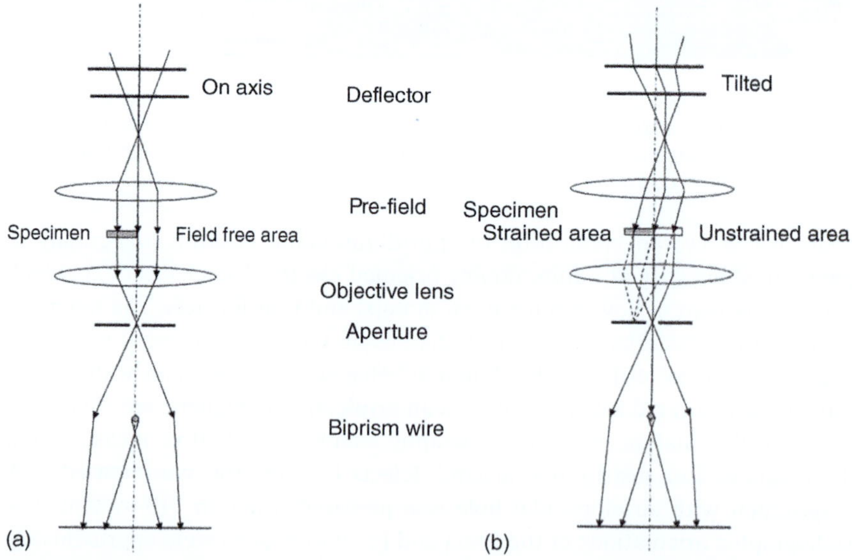

Figure 6.3 Ray path diagram for (a) conventional electron holography and (b) dark-field electron holography.

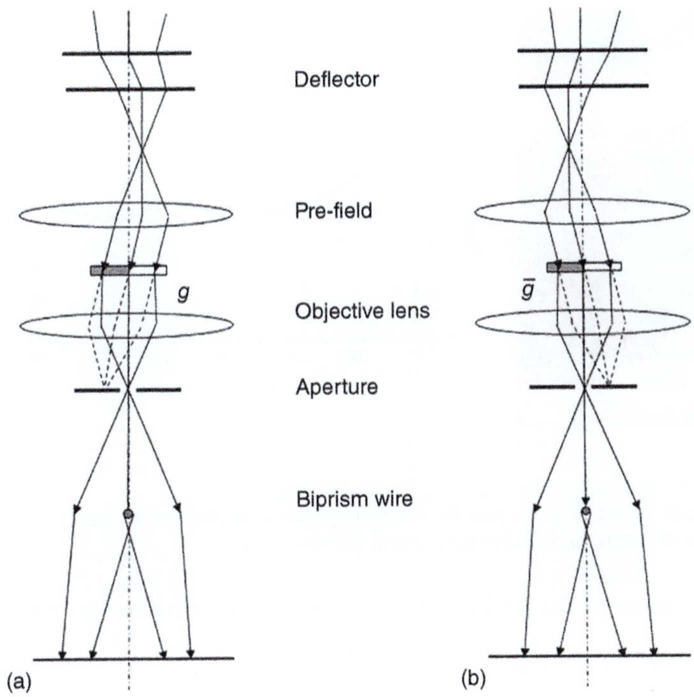

Figure 6.4 (a) g-reflection ray path diagram and (b) $\bar{g}$-reflection ray path diagram for electron holography.

in similar diffraction conditions. This geometry resembles most cross-sectional specimens of semiconductor thin layers or devices grown epitaxially on a substrate. The two diffracted beams can then be interfered with the aid of an electrostatic biprism. Their phase difference, which can be measured in two dimensions directly from the holographic fringes, depends on the dynamical elastic scattering and, more importantly, on the geometric phase as defined by geometric phase analysis. If the specimen is of uniform thickness, the former will be a constant phase term while the latter encodes the strain information through phase gradients. With this technique, the deformation of a strained silicon transistor array has been accurately measured.

This technique was further extended and applied to strain detection around the precipitates in a magnetic material. For magnetic materials, the strain field information should be separated from the magnetic information [13]. Thus, as shown in Figure 6.4, two sets of interference fringes are observed under dark-field conditions with Bragg reflections g and $\bar{g}$ excited.

Figure 6.5a clarifies several essential features of the lattice strain in the $Nd_2Fe_{14}B$ phase including the α-Nd precipitates. The color bar at the bottom shows the magnitude and sense of the lattice deformation. First, it appears that the $Nd_2Fe_{14}B$ lattice (i.e. the spacing of the c planes) is elongated to the left and right sides of the precipitates. These sides are defined with reference to the direction of the c axis, indicated by the arrow in the figure. Above and below the precipitates, the lattice

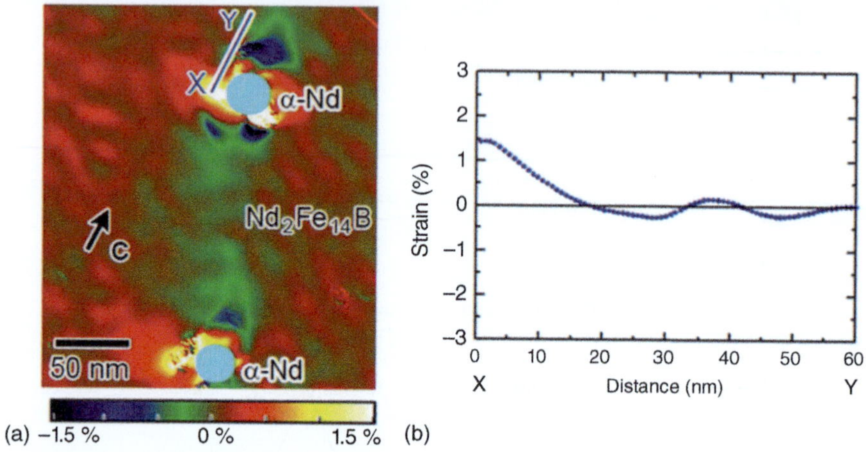

Figure 6.5 Strain map determined by DFEH. (a) Strain map representing elongation/compression of spacing of c planes in $Nd_2Fe_{14}B$ lattice. Color bar indicates sense and magnitude of strain. (b) Plot of strain along line X–Y in (a).

is compressed. The sense of the strain is thus complementary depending on the observed region. Figure 6.5b plots the strain values measured along the X–Y line in (a). The strain is seen to gradually increase as the measurement position moves closer to the precipitate. The maximum strain (approximately 1%) was observed in the vicinity of the precipitate.

6.3 Lorentz Microscopy

As noted above, electron holography can visualize electromagnetic fields both inside and outside a specimen. In this section, we describe a technique using a transmission electron microscope that can be used to visualize domain structures and/or domain walls. Since the principles of this imaging technique are understood in terms of Lorentz force on the incident electrons, as explained below, the imaging technique is called "Lorentz microscopy."

In electron holography, the interference effect of incident electron is used, and the spatial coherence length of the incident electron is given by $\lambda/2\alpha_{\mathrm{i}}$, where λ is the wavelength of the electron and α_i is the illumination angle of the incident electron beam (see Section 5.3). For obtaining highly coherent condition, illumination angle should be small, which means that collimated illumination is necessary. Hence, the basic principles of the imaging process of electron holography can be formulated accordingly via quantum mechanics.

In contrast, an electromagnetic field can be detected and visualized by considering the Lorentz force on incident electrons. Lorentz microscopy observation can be understood via the classical particle picture of incident electrons. The illumination angle of the incident electron beam in Lorentz microscopy can be larger than that in electron holography, so a more intense electron beam can be used. This technique is thus useful for dynamic observation of electromagnetic fields with a video system [14, 15].

6.3.1 Fresnel Mode (Defocusing Mode)

The Lorentz force $\vec{F}$ for an electron passing through a specimen with magnetic flux density $\vec{B}$ (T) is given by

$$\vec{F} = -e(\vec{v} \times \vec{B}) \tag{6.1}$$

where $\vec{v}$ is a vector of the electron velocity and e is the elementary electric charge. The direction of the force expressed by the vector product in Eq. (6.1) is given by Fleming's left-hand rule, so the trajectory of the incident electrons is as shown in Figure 6.6a. Here, we assume that the 180° domains are magnetized perpendicularly to the plane of the diagram and in opposite directions in alternate domains, as shown in (b). The electron beam is thus deflected in opposite directions in adjacent domains, resulting in alternating deficiency and excess of electrons in the region below the lower specimen surface. If we observe the specimen in the over-focus condition, increases and decreases in the electron intensity are observed periodically at domain wall positions, as shown in Figure 6.7c. On the other hand, if we observe the specimen in the under-focus condition, the image contrast is reversed, as shown in Figure 6.6d.

Figure 6.7a,b shows Lorentz microscopy images of an as-quenched $Fe_{73.5}Cu_1Nb_3Si_{13.5}B_9$ specimen observed in under- and over-focus conditions, respectively. These

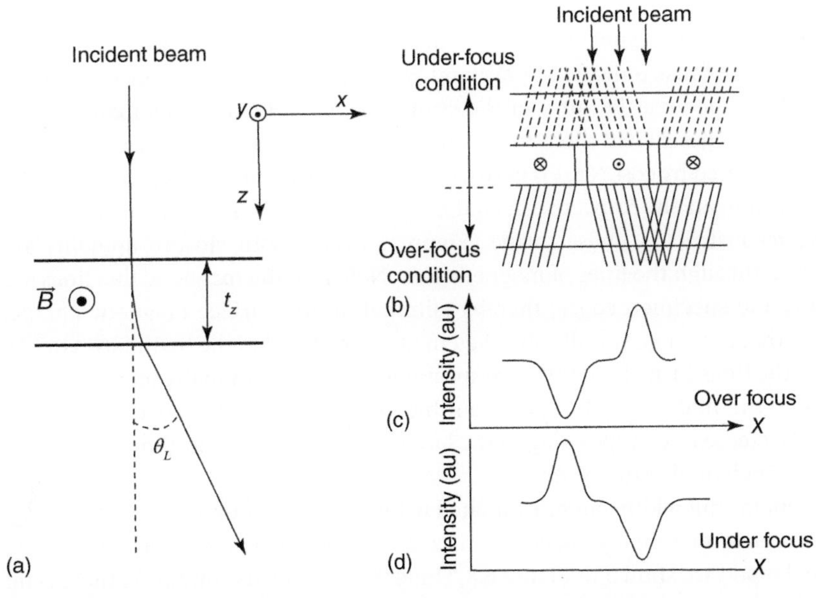

Figure 6.6 (a) Trajectory of electron passing through magnetized specimen. (b) Schematic of Lorentz microscopy in Fresnel mode. Intensity distributions for (c) over- and (d) under-focus conditions.

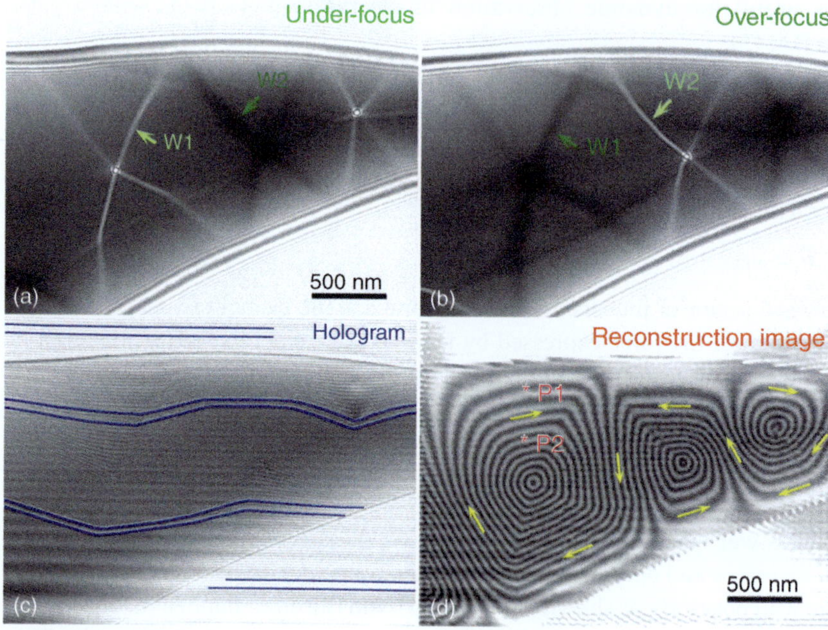

Figure 6.7 Lorentz microscopy images of as-quenched $Fe_{73.5}Cu_1Nb_3Si_{13.5}B_9$ specimen observed in (a) under- and (b) over-focus conditions. (c) Electron hologram of specimen. (d) Reconstructed phase image of specimen (arrows indicate direction of lines of magnetic flux). Source: Shindo and Akase [1], with permission from Elsevier.

images, observed in Fresnel mode, show magnetic domain boundaries as white and black bands like those indicated by W1 and W2. Figure 6.7c shows an electron hologram of the same area. Because of the strong magnetic field of this material, the interference fringes switch direction from time to time. In the reconstructed phase image represented by $\cos(\phi(x, y))$ in (d), the densities and directions (arrows) of the white lines indicate the respective densities and directions of the magnetic flux lines.

In the reconstructed phase image (Figure 6.7d), smooth closure domains are clearly seen through the lines of magnetic flux. Note that the magnetic flux lines are parallel to the specimen edges, thereby eliminating the surface magnetic charge. Also note that the domain walls observed in (a) correspond to the boundaries in (d), in which the lines turn about 90°. The specimen thickness gradually increases from the edge to the middle; as the specimen thickness increases, the spacing between the lines decreases, corresponding to the larger magnetic flux accordingly projected along the incident electron beam.

The domain wall width can be evaluated using images captured in Fresnel mode. Typical Lorentz microscopy images of an as-sintered Sm–Co magnet observed using in Fresnel mode are shown in Figure 6.8. Under the just-focus condition, there is no definite image contrast in (a), whereas under the defocus condition, white or black lines appear at the positions of the magnetic domain walls depending on the defocus sign (b) and (c). The width of the white or black lines directly depends on the focus

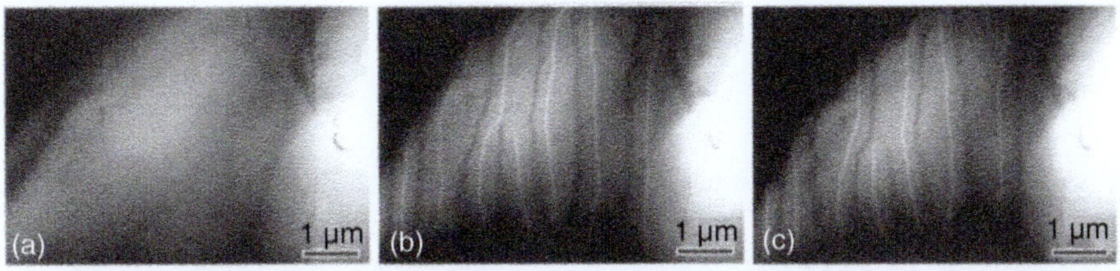

Figure 6.8 Lorentz microscopy images of as-sintered Sm–Co magnet observed using transmission electron microscope in Fresnel mode under three conditions: (a) just-focused, (b) under-focused, and (c) over-focused.

setting, and the wall width can be evaluated quantitatively by taking into account the experimental conditions such as the focus setting.

The domain wall width in magnetic materials can be determined by using a linear extrapolation method. According to geometric-optical theory [16], a full width W_d (Δf) at half maximum (FWHM) of a divergent wall image at defocus value Δf is given by [17]

$$W_d\,(\Delta f) = 2\,x_{1/2}(\Delta f) + 2\,\Delta f\,\theta_L\,\frac{\Phi_y(x_{1/2})}{\Phi} \tag{6.2}$$

where Φ is the magnetic flux inside the specimen, $\Phi_y(x)$ is the y component of the magnetic flux, and θ_L is the Lorentz deflection angle. The coordinates x, y, and z are indicated in Figure 6.9a. In (b), wall width 2δ is defined by the slope of function $d\Phi_y(x)/dx$ at the wall center, and $2x_{1/2}$ is given as the FWHM of the function $d\Phi_y(x)/dx$, as shown in (c). Therefore, by extrapolating the FWHMs (W_d) of divergent wall images captured at various defocus values, the domain wall width at $\Delta f = 0$ can be experimentally determined. The domain wall width obtained in this way was

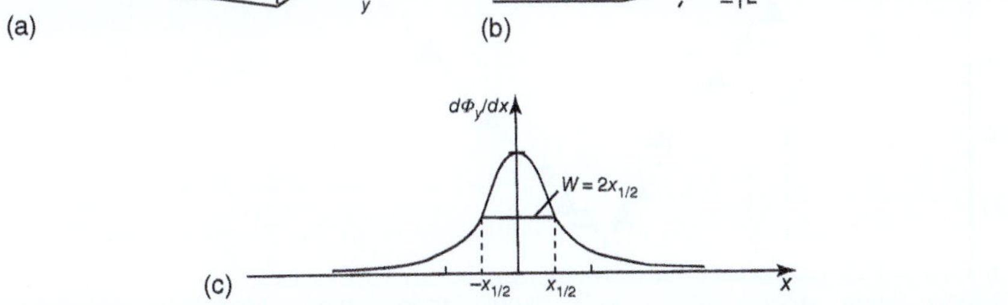

Figure 6.9 (a) Schematic illustration of 180° domain wall in film with uniaxial anisotropy. (b) Wall width 2δ defined by slope of function $d\Phi_y(x)/dx$ at wall center. (c) $2x_{1/2}$ given as FWHM of function $d\Phi_y(x)/dx$.

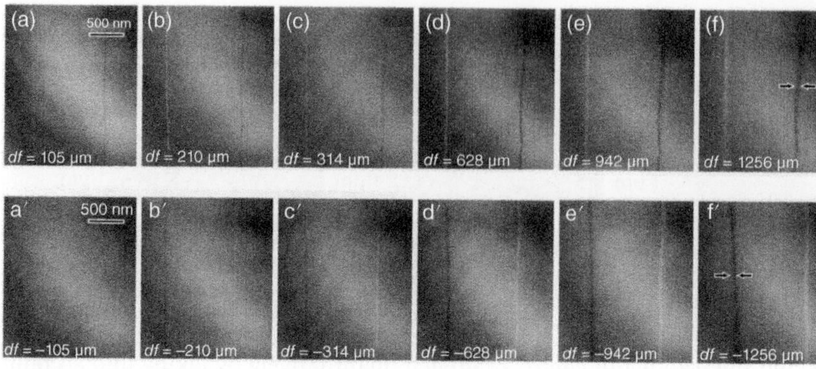

Figure 6.10 Lorentz microscopy images of as-sintered Sm–Co magnet for series of focus changes.

shown to be not significantly affected by illumination angle α_i of the incident beam provided that α_i is of the order of θ_L or less [18, 19].

Figure 6.10 shows a series of Lorentz microscopy images observed in Fresnel mode for an as-sintered Sm–Co magnet. The images were observed systematically while changing the defocus value. The wall widths were measured from the intensity profiles of divergent wall images recorded on imaging plates. These profiles were obtained in the narrowest areas of the divergent wall images to avoid overestimation of the wall width due to tilting of the domain wall against the electron beam. The FWHM of a wall image plotted against the defocus value is on a straight line, as shown in Figure 6.11. The domain wall width of an as-sintered

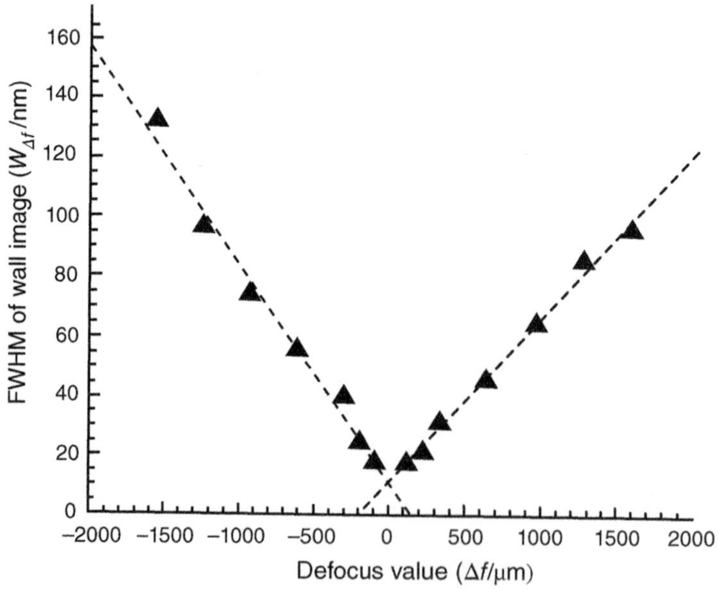

Figure 6.11 FWHM of wall image measured as function of defocus value.

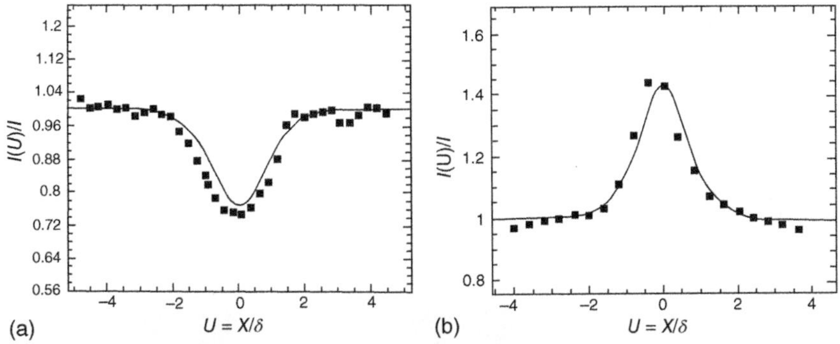

Figure 6.12 Observed (squares) and theoretical (curves) intensity distributions of domain wall images for as-sintered Sm–Co magnet against calculated intensity distributions for (a) divergent and (b) convergent cases. Defocus value is 0.63 mm.

Sm–Co magnet was evaluated to be 10 nm [20]. This value is consistent with the assumption for evaluation of the coercive force of an Sm–Co magnet [21, 22].

Figure 6.12 shows the intensity distributions of the domain wall images in an as-sintered Sm–Co magnet observed by Fresnel mode against the calculated intensity distributions under divergent and convergent conditions. The squares and curves represent the observed and theoretical intensity distributions, respectively, of the domain wall. By considering the spin distribution across the wall region for the minimum total energy resulting from exchange and either anisotropy or magnetostatic energy [23], the equation of the theoretical intensity distribution of the domain wall images [24, 25] is derived as

$$I(U)/I = |1 \pm R\,\mathrm{sech}^2\, u|^{-1}, U = u \pm R\ \tanh\, u \tag{6.3}$$

where $u = x/\delta$, $U = X/\delta$, $\mathrm{R} = \Delta f\ \theta_L/\delta$, δ is the half of the domain wall width and x and X are the coordinates normal to the wall direction in the specimen and image planes, respectively. $I(U)$ is the electron beam intensity in the image plane; I is the uniform illuminating intensity. The plus and minus signs in the equation correspond to the divergent and convergent cases, respectively. The equation indicates that the width of the white lines in the convergent images is substantially less than that of the black lines in the divergent images. As shown in Figure 6.12, the experimental intensity profiles measured with the imaging plate at a defocus value 0.63 mm were best fitted with the calculated intensity profiles in the divergent and convergent cases at $R = 0.3$ [26].

6.3.2 Foucault Mode (In-Focus Mode)

In the electron diffraction pattern obtained from an area containing a domain wall separating two magnetic domains, as shown in Figure 6.13, the diffraction spots split and form two spots. This splitting can be observed directly in the electron diffraction pattern, for both the transmitted and diffracted beams. The Lorentz

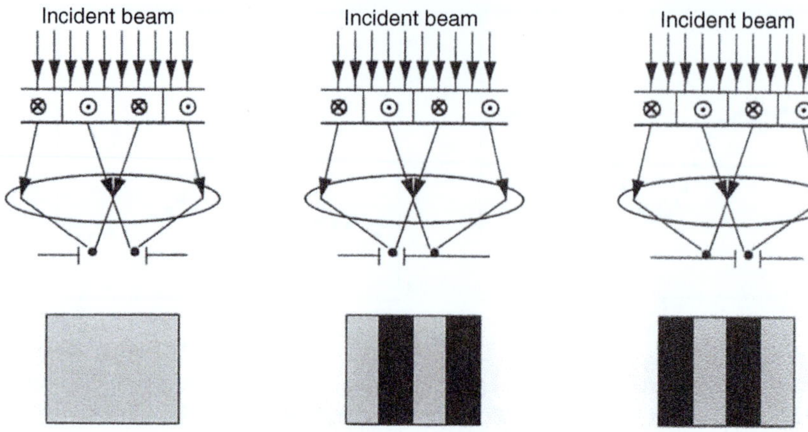

Figure 6.13 Schematic illustration of Lorentz microscopy in Foucault mode.

force on an electron passing through a magnetized specimen with magnetization $\vec{I_s}$ ($A\,m^{-1}$) is given by

$$\vec{F} = -e\mu_0(\vec{v} \times \vec{I_s}) \tag{6.4}$$

where μ_0 ($H\,m^{-1}$) is the permeability in vacuum. The direction of the magnetization in a specimen of thickness t_z is assumed to be normal to the plane of the diagram (i.e. the y-direction), as shown in Figure 6.6. The electron is deflected from the straight trajectory due to magnetization $\vec{I_s}$, but its energy remains constant. From Eq. (6.4), the Lorentz force on the incident electron, which is parallel to the x-direction, is expressed as

$$\frac{dp_x}{dt} = m'\frac{d^2x}{dt^2} = ev\mu_0 \tag{6.5}$$

where m' is the electron mass at velocity v:

$$m' = m\Big/\left(1 - \frac{v^2}{c^2}\right)^{\frac{1}{2}}, \tag{6.6}$$

with m being the rest mass and c being the speed of light. In Eq. (6.5), dp_x/dt is the x component of the Lorentz force, giving $p_x = m'(dx/dt) = et_z\mu_0 I_y$, because the electron's velocity can be treated as a constant. For the magnitude of the magnetization given by $I_y = I_s$ and $I_x = I_z = 0$, deflection angle θ_L is given by

$$\theta_L = \frac{dx}{dz} = \frac{dx}{dt}\frac{dt}{dz} = \frac{et_z}{m'v}\mu_0 I_y \tag{6.7}$$

The angle of deflection by the Lorentz force is rather small compared with the Bragg angle. For instance, deflection angle θ_L for $Nd_2Fe_{14}B$ with a thickness of 100 nm and $B = \mu_0 I_s = 1.6\,T$ is 3×10^{-5} rad for 1250-keV electrons. The electron diffraction pattern in Figure 6.14a, obtained from a sintered $Nd_2Fe_{14}B$ magnet, shows the splitting of each spot. The separation of the spots was ~5×10^{-5} rad,

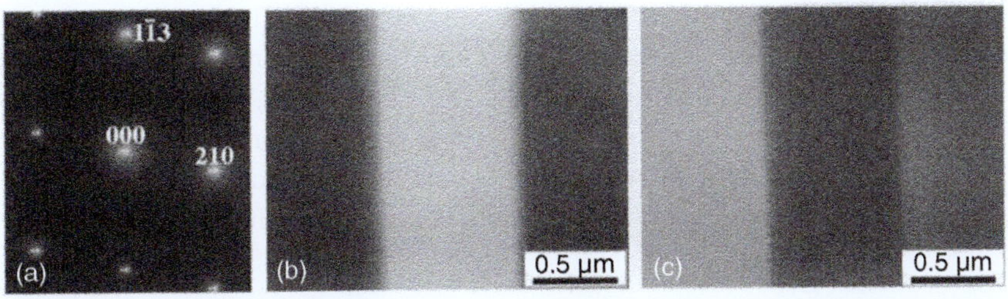

Figure 6.14 (a) Electron diffraction pattern of sintered $Nd_2Fe_{14}B$ magnet. (b and c) Typical Lorentz microscopy images of magnetic domains observed in Foucault mode when objective aperture eliminated one of the split transmitted beams.

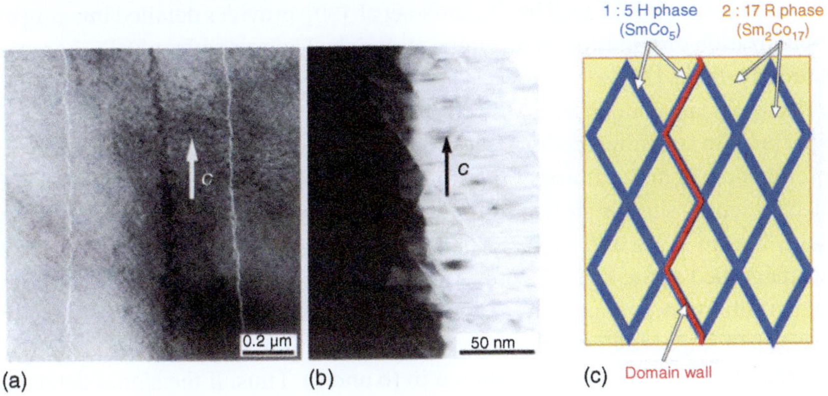

Figure 6.15 Lorentz microscopy images of step-aged Sm–Co magnet observed in (a) Fresnel mode and (b) Foucault mode. Arrow in (a) indicates [001] direction. (c) Schematic illustration of geometrical relationship between domain wall (along $SmCo_5$ phase) and specimen microstructure.

which is consistent with the estimation. By displacing the objective aperture in the back focal plane, it is possible to exclude one of the two transmitted beams. As a result, alternate magnetic domains appear periodically as dark and bright regions, as shown in (b) and (c).

Figure 6.15a,b shows Lorentz microscopy images of a step-aged Sm–Co magnet observed in Fresnel and Foucault modes, respectively. Whereas the image in (a) shows only serrated domain walls, indicating pinning at the cell boundaries, the one in (b) shows not only the magnetic domain but also the microstructure. The microstructure consists of 1 : 5 H and 2 : 17 R phases, where the 1 : 5 H phases of the cell boundaries act as attractive pinning centers for the magnetic domain walls in the step-aged magnet, as schematically illustrated in (c). It can be seen that domain wall pinning increases greatly with an increase in the domain wall energy gradient between the 1 : 5 H and 2 : 17 R phases, owing to the chemical partitioning of the additives of Cu and Fe atoms by step aging [15, 27].

A modified observation mode for the Foucault mode has been developed [28]. In this mode, half the diffraction pattern is eliminated with a phase-shifting aperture, enabling the interference contrast in the domain to be obtained. The aperture itself is one of three types: opaque half-plane, phase-shifting half-plane, or phase-shifting small hole. The mode is called the "coherent Foucault mode." If the edge of the phase-shifting aperture is put at the center of a transmitted beam, interference contrast is observed in all directions. A coherent Foucault image is obtained in real time without any postprocessing, so using this mode is useful for in situ experiments [29].

6.3.3 Lorentz Microscopy Using Scanning Transmission Electron Microscope

Like electron holography described in Section 4.2, differential phase contrast (DPC) Lorentz microscopy, introduced by Chapman et al. [30], provides detailed imaging of the magnetization distribution in magnetic materials. The principles of DPC Lorentz STEM are as follows [31].

The electron beam condensed by the objective pre-field lens is focused on the magnetic specimen in scanning mode. At each point, because of the beam's deflection by the Lorentz force, a disk pattern of the transmitted beam in the back focal plane of the objective lens is slightly shifted from the center of the ray axis. Figure 6.15 shows how, with intermediate and projector lenses, the disk on the objective back focal plane is magnified at the split detector. When the convergent beam passes through the magnetic domain, as shown in (a), the deflection disk is shifted to the right of the detector, and the signal detected by detector (B + C) is stronger than that detected by detector (A + D), and vice versa, as shown in (b and c). Thus, if the signal detected by (B + C) is subtracted from that detected by (A + D), the signal component due to the phase shift is enhanced while a component not due to the phase shift vanishes because the latter component is detected equally by both detectors. In this way, the magnetization direction can be detected.

Figure 6.16 shows a schematic diagram of a quadrant-detector imaging system [32, 33]. By adding or subtracting each signal, domains in every direction can be displayed without rotating the specimen, as shown in Figure 6.17.

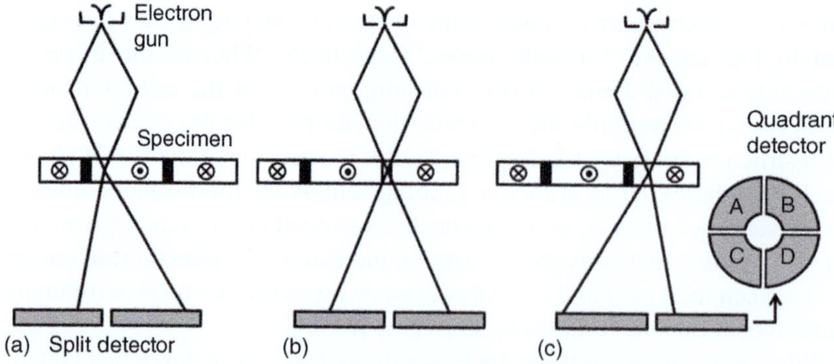

Figure 6.16 Geometric configuration of electron gun and quadrant detector in DPC Lorentz microscopy.

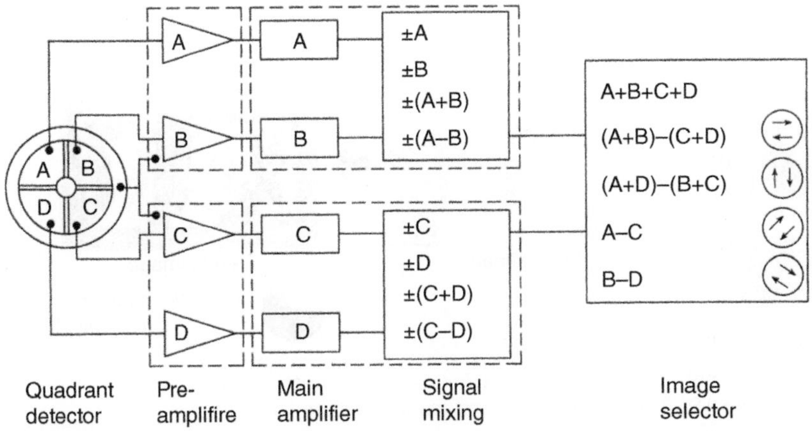

Figure 6.17 Geometric configuration of quadrant detector and image processing system in DPC Lorentz STEM.

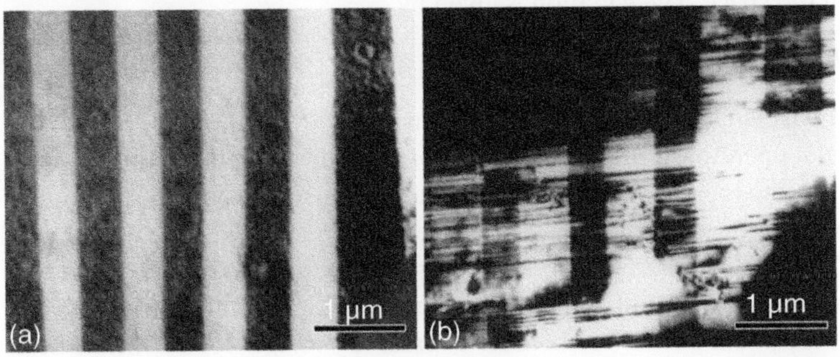

Figure 6.18 Images of Co polycrystal captured by (a) DPC Lorentz STEM and (b) Lorentz TEM in Foucault mode. Source: Images were kindly provided by Dr. K. Tsuno.

Figure 6.18 shows magnetic domain images of a Co polycrystal captured by DPC Lorentz STEM (a) and Lorentz microscopy in Foucault mode (b). The advantage of DPC Lorentz STEM over conventional Lorentz microscopy is apparent. The change in image contrast due to specimen thickness variation evident in (b) is not evident in (a). Imaging of slip bands in the horizontal direction can also be removed by careful adjustment of the detector position [34]. Thus, image contrast corresponding to the magnetic domain structure can clearly be extracted by using DPC Lorentz STEM.

DPC Lorentz STEM has recently been performed with a probe size of less than 0.1 nm for clarifying the electric field between atomic columns [35]. It has also been used to determine the three-dimensional (3D) electrostatic field at an electron nano-emitter [36].

6.3.4 Phase Reconstruction Using Transport-of-Intensity Equation

From a defocus series of Lorentz microscopy images observed in Fresnel mode, the phase information of an electron beam can be derived by using the

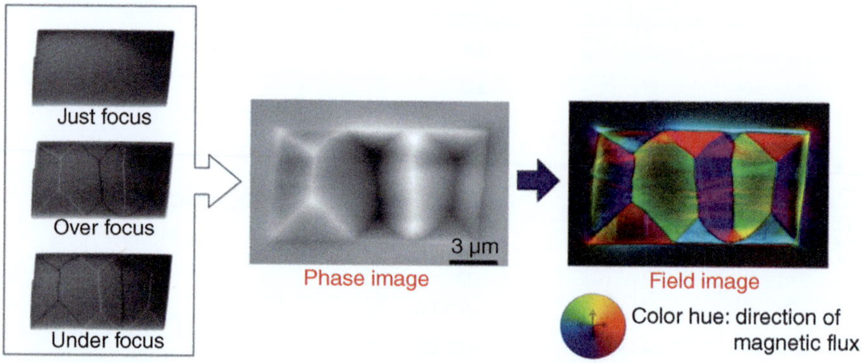

Figure 6.19 Example of phase reconstruction using TIE equation for $Fe_{18.8}Co_{60}Cu_{0.6}Nb_{2.6}Si_9B_9$ nanocrystalline soft magnetic material.

transport-of-intensity equation (TIE) [37–42]. This equation describes the internal relationship between the intensity and phase distribution when a wave is propagating in the space. The equation is given by

$$\frac{2\pi}{\lambda}\frac{\partial}{\partial z}I(x,y,z) = -\nabla_{xy} \cdot (I(x,y,z)\nabla_{xy}\phi(x,y,z)) \tag{6.8}$$

where λ is the wavelength of the incident electron beam, $I(x, y, z)$ is its intensity, ∇_{xy} is the differential operator in the x–y plane, and $\phi(x, y, z)$ is the phase shift. Paganin and Nugent suggested solving Eq. (6.8) by using a Poisson equation [43]. Solving the phase shift requires partial differentiation of the beam intensity in the direction of propagation. This partial differentiation can be approximated via divided differences of Lorentz microscopy images observed in Fresnel mode. Specifically, the phase shift image can be calculated from three Lorentz microscopy images for which the defocus values are symmetric around the in-focus value [44, 45].

Figure 6.19 shows an example of this phase reconstruction using the TIE method. The specimen is an $Fe_{18.8}Co_{60}Cu_{0.6}Nb_{2.6}Si_9B_9$ nanocrystalline soft magnetic material thinned by FIB milling. The left panel shows the three Lorentz microscopy images. From these images, the phase image in the center panel was calculated using Eq. (6.8). The magnetic field image shown in the right panel, in which the magnetic flux is mapped by color hues, was converted from the phase image because the specimen's thickness was almost uniform. The accuracy of the phase obtained using the TIE method is easily affected by low-frequency noise. Nevertheless, this method is useful for analyzing magnetic domain structures because it does not require any special equipment for the transmission electron microscope and does impose any limitations on the field of view.

6.4 Magnetically Shielded Lens and High-Voltage Electron Microscope

Observation of the magnetic domain structure in magnetic materials by TEM requires special care because the magnetic field at the specimen position in a 200- or 300-kV transmission electron microscope is about 1.8–2.0 T, which is strong enough to destroy or modify the inherent magnetic domain structure. Thus, the magnetic field at the specimen position should be reduced, especially for observing soft magnetic materials. One of the easiest ways to reduce the magnetic field is to switch off and degauss the objective lens. Doing this can reduce the residual magnetic field to less than 0.2 mT.

However, fine TEM images cannot be obtained if the objective lens is switched off because the objective lens determines the resolution of a transmission electron microscope. To maintain the advantage of using the objective lens for observing detailed magnetic structures, the specimen should be positioned near the peak of the magnetic field of the lens. This is because, to get good quality images, the lens field must be near the specimen so that the electrons scattered by the specimen can be immediately collected near the optical axis.

The objective lens pole pieces are thus shaped as shown in Figure 6.20. To maintain the action of the objective lens for observing detailed magnetic structure, the position of the specimen should be shifted out of strong magnetic field, or special magnetic shielding should be introduced in the objective lens. Figure 6.20 shows cross sections of single-gap (a) and double-gap (b) lens systems for a magnetic field-free objective lens used to observe magnetic domain structures. Figure 6.21 shows the simulated

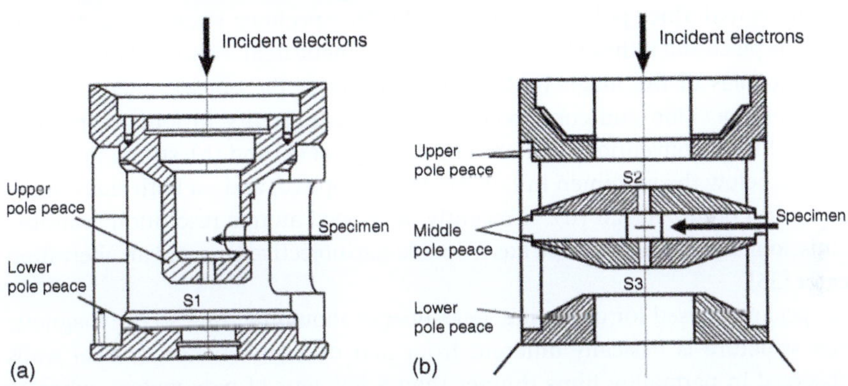

Figure 6.20 Schematic illustrations (cross-sectional views) of magnetically shielded objective lens pole pieces with (a) a single gap (S_1) and (b) a double gap (S_2, S_3).

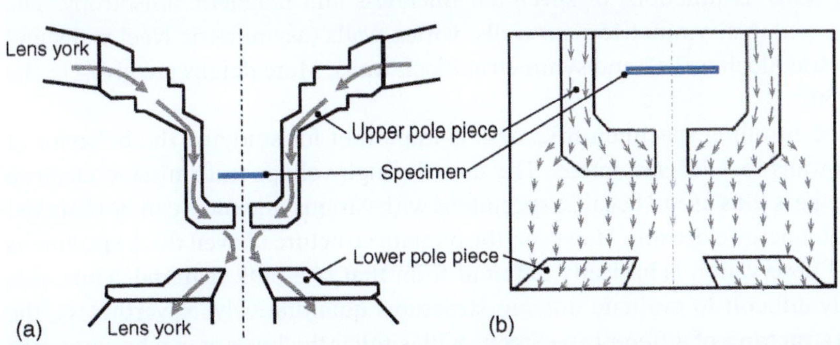

Figure 6.21 (a) Simulated magnetic field around specimen and lens gap obtained using 3D model. (b) Enlarged image at lower part of (a).

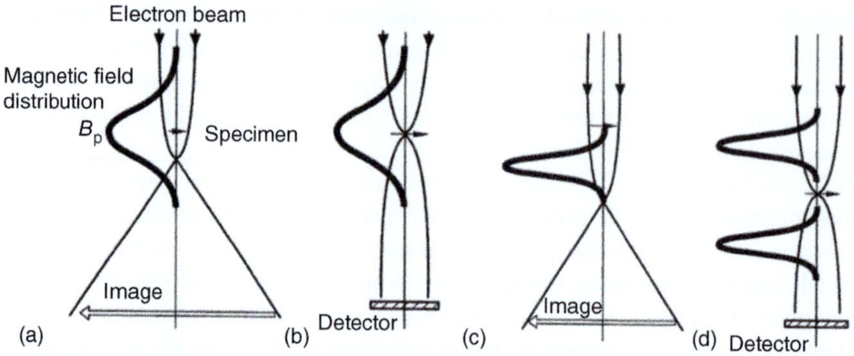

Figure 6.22 Schematic illustration of magnetic field distribution of objective lens and corresponding electron trajectory for (a) conventional TEM, (b) conventional STEM, (c) Lorentz microscopy, and (d) Lorentz STEM.

magnetic field indicated by arrows around the specimen and lens gap obtained using a 3D model for the single-gap lens system.

The details of a double-gap objective lens were reported [34, 46] and used for DPC Lorentz STEM (see Section 6.3). The overall pole-piece structure has two gaps, S_2 and S_3, along with three pole pieces and a hole. The specimen is surrounded by the middle pole piece and is thus shielded from the magnetic field. Figure 6.22 illustrates typical examples of the magnetic field distribution of the objective lens around the specimen position and corresponding electron trajectory in four microscopy modes. Unlike in conventional electron microscopy of (a) and (b), strong magnetic fields form below the specimen in Lorentz microscopy (c) and on both sides of the specimen in Lorentz STEM (d). A recently developed atomic-resolution scanning transmission electron microscope has a double-gap objective lens and an aberration corrector [35].

The specimens used for observing TEM images should be thin, so the magnetic domain structure is basically different from that of a bulk material. Néel walls are observed in permalloy films thinner than a few tens of nanometers, whereas cross-tie walls with a ripple structure are observed in films thicker than several tens of nanometers [47, 48]. Ramstök et al. [49] reported a phase diagram of domain walls as functions of specimen thickness and magnetic anisotropy. The diagram contains symmetric Néel walls, vortex walls (asymmetric Néel walls and asymmetric Bloch walls), and symmetric Bloch walls. More details are given in the report [50].

Thus, controlling specimen thickness is important for studying the behavior of domain walls in TEM specimens. The use of a high-voltage transmission electron microscope is thus useful because specimens with various thickness can be observed with high-energy electrons. However, the domain structure of even thick specimens for TEM observation is basically different from that of a bulk material. Thus, it is generally difficult to evaluate domain structures quantitatively. Nevertheless, the domain structures of different specimens with similar thicknesses can be compared to systematically understand the differences in their magnetic properties. On the

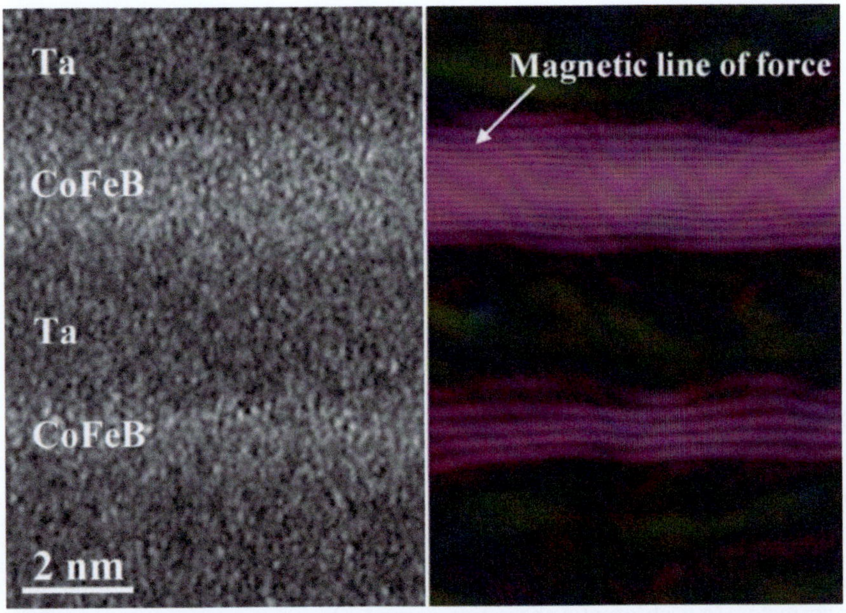

Figure 6.23 TEM image (left) and magnetic flux distribution (right) of CoFeB/Ta multilayer. Magnetic flux represented by cosine of phase φ_M amplified 600 times (cos $600\varphi_M$) with smoothing over length scale of 1.43 nm parallel to CoFeB layer. Constant flux of h/600e flowed between adjacent contour lines.

other hand, the domain structures of ultrathin magnetic layers and nanoparticles can be interpreted directly in relation to their magnetic properties [51–56].

A high-voltage transmission electron microscope has various other advantages. As described in Section 8.4, it can be used to perform three-dimensional magnetic structure analysis. To obtain many reconstructed phase images projected in various directions, the specimen should be tilted at large angles, causing the specimen's relative thickness to change accordingly. The use of high transmission capability is thus effective for such analyses. A high-voltage microscope can also be used to effectively confirm the magnetic structure with a change in specimen thickness, as in the case of a skyrmion lattice, as explained in Section 8.2.3. The recent development of an aberration corrector for the objective lens system has enabled high-voltage, high-resolution holography TEM. This means that the detailed magnetic structure can be analyzed with a resolution of less than 1 nm, as evidenced by the TEM image and magnetic flux distribution of a CoFeB/Ta multilayer shown in Figure 6.23 [57].

6.5 Aberration-Corrected Lens System

In this section we first describe the factors that cause aberrations and the types of aberrations that occur in transmission electron microscopes. Next, we discuss the aberrations that need to be carefully treated depending on the magnification range.

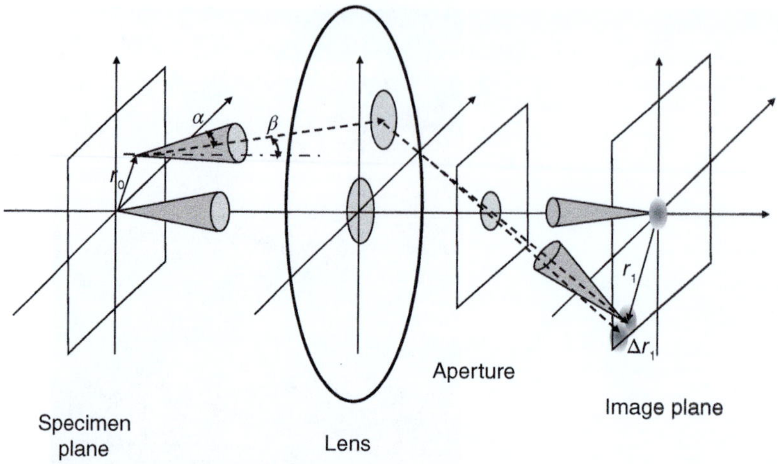

Figure 6.24 Schematic illustration of effect of lens aberration on path of scattered electrons.

Finally, we note the on-axis aberrations that remain (e.g. spherical aberration) and their correctors (e.g. chromatic aberration corrector).

As shown in Figure 6.24, electrons start from a point on the object plane with some spreading angle. Therefore, when the incident electrons pass through a magnetic lens, they converge to a Gauss image point on the image plane. However, if the imaging process is not perfect, the convergence point will not match the Gauss image point, so each point will be blurred.

Table 6.1 shows the magnitude of various lens aberrations that can occur in the imaging system. At low magnification, the trajectory of the electron beam might not pass along the axis of the projector lens, resulting in distortion aberrations such as spiral, pincushion, and barrel-shaped aberrations. The magnitude of the distortion aberration is proportional to the cube of the distance from the optical axis. Among these aberrations, the spiral distortion is unique about the magnetic lens; it results in

Table 6.1 Aberrations of objective, intermediate, and projector lenses.

	Objective lens	Intermediate lens	Projector lens
Spherical aberration	◎	X	X
Coma	△	X	X
Curvature of field	△	◎	△
Astigmatism	△	◎	△
Distortion	X	○	◎
On axis chromatic aberration	◎	X	X
Magnification chromatic aberration	△	◎	△
Rotational chromatic aberration	△	△	△

◎, large; ○, not small; △, small; X, negligible.

rotational motion of the electrons in the lens field. To balance the lens polarity, this distortion can be suppressed and ignored. These aberrations are remarkable when observing low-magnification images.

The distortions and blurs caused by a magnetic lens depend on the lens field, the optical axis angle, and the aperture opening angle. To adjust the beam tilt and opening angle (Figure 6.24), which determine the aberration magnitude for each lens, a deflection system is incorporated in the illumination system (shown in Figure 5.2) and used to adjust the beam position.

When observing electron microscope images, the magnification is determined by the distance between the object plane and image plane for each lens. In general, in the final image, the contribution of aberration from each lens should be set sufficiently small. At medium magnification, the electron beam passes through the intermediate lens off-axis, resulting in off-axis chromatic aberration and field curvature. At high magnification, the electron beam used for the image formation in the objective lens is almost on axis and paraxial, so the off-axis aberrations are negligible compared with the magnitudes of the aberrations at medium and low magnifications and thus can be ignored. In a properly aligned microscope, spherical, axial chromatic, and diffractive aberrations remain as the final amount of image blur.

To reduce the spherical aberration coefficient of the objective lens, the half width of the magnetic field distribution on the optical axis must be reduced, and the peak value of the magnetic field must be increased. Although the magnitude of the aberration can be reduced in this way, it cannot be reduced to zero with a rotationally symmetric magnetic field lens. As shown in Figure 6.25, concave

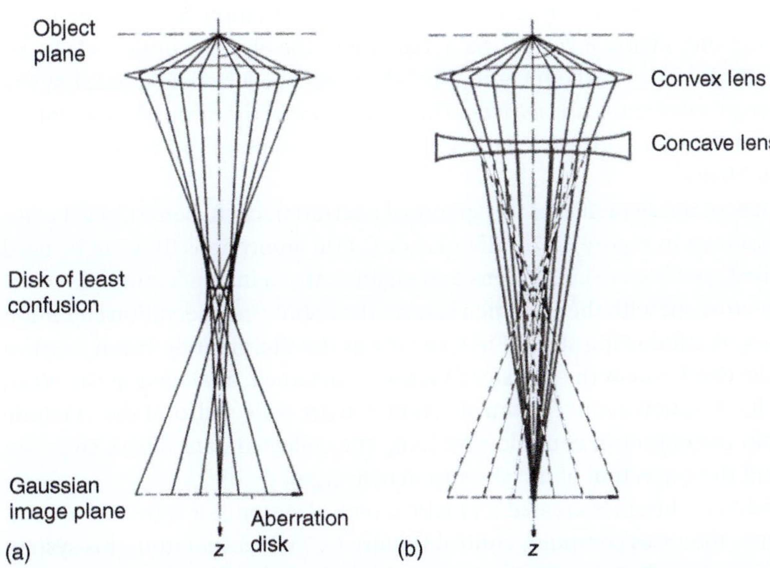

Figure 6.25 (a) With convex lens, electron beam spreads over spherical aberration disk at Gaussian image plane. (b) With concave lens, ray path can be corrected so that all rays converge to a point on image plane.

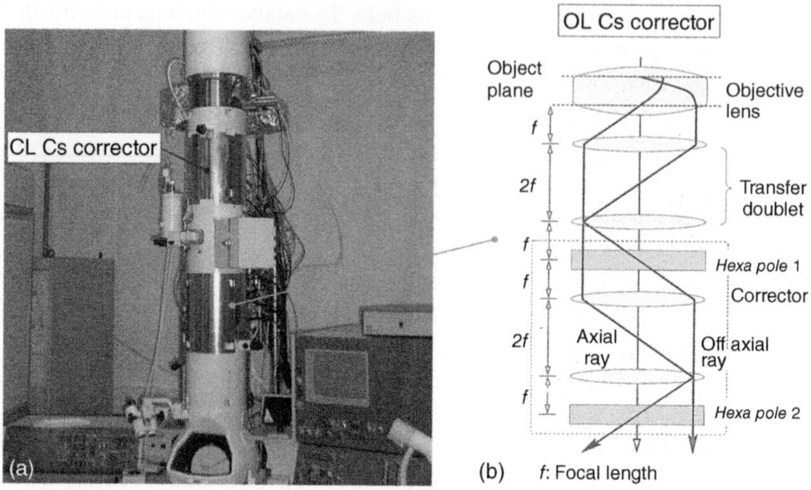

Figure 6.26 (a) 200-kV transmission electron microscope with spherical aberration hexapole correctors incorporated into illumination and imaging sites. (b) Diagram of electron optical elements of spherical aberration hexapole corrector installed at imaging site.

lens action is required to eliminate spherical aberration, but due to the accuracy and stability, it is practically difficult to control the lens action. Therefore, this aberration must be controlled using a corrector system comprising two hexapole correctors.

Figure 6.26a shows an outer view of a 200-kV transmission electron microscope incorporating two hexapole correctors, one at the illumination site and the other at the imaging site. Figure 6.26b shows a diagram of the electron optical elements in the corrector installed at the imaging site. The magnetic lens field created by the hexapoles is equivalent to a concave lens. The aberrations of the hexapoles can cancel each other out, and the concave lens action with the hexapoles can be balanced by using transfer lenses.

An overview of the sequence of the spherical aberration coefficient (*Cs*) collector operation is shown in Figure 6.27. Either Ge or C thin amorphous film can be used as the standard specimen (a). After lens axis alignment at a magnification of several ×10 k, the microscope with the specimen is set to the optimum observation condition by performing the following steps. First, the tilt of the illumination beam is set in order, and the condition of the transfer function is obtained, as shown in (b). Next, tableaux of information about the transfer function are collected, and the machine is again set to the optimum condition by using the collected data. These steps are repeated until the correction of the aberration converges.

Haider and his colleagues created a practical optical system for spherical aberration correction that uses computer control (Figure 6.27). Incorporating this system into a transmission electron microscope enables the spherical aberration to be corrected.

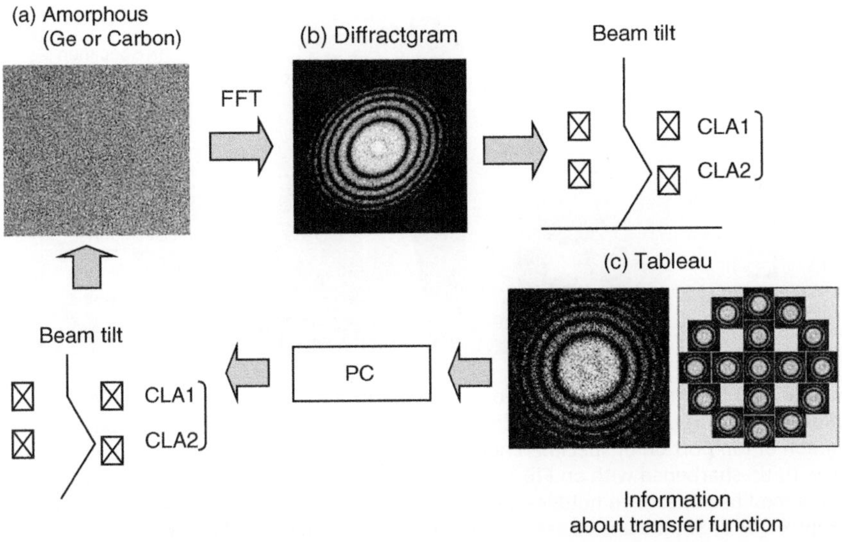

Figure 6.27 Spherical aberration correction process. (a) Highly magnified image of amorphous Ge. (b) Corresponding digital diffractogram. (c) Diffractogram tableau obtained by automatically tilting electron beam.

For the remaining aberration, i.e. axial chromatic aberration, a correction device has been designed for practical use. The principle of the correction is based on the fact that the magnetic force increases with the velocity of the charged particles while the electrostatic force remains constant. A prototype axial spherical and chromatic aberration corrector has been constructed and tested [58].

6.6 Multifunctional Specimen Holders with Piezodriving Probes

Figure 6.28a shows a schematic diagram of a double-probe piezodriving holder [59]. A micrograph of the top portion of this specimen holder is shown in (b). There are two arms for mounting two metallic probes. The probes can be manipulated independently in three dimensions by the motion of arms 1 and 2, which are driven by micrometers and/or piezoelectric elements. Each probe has three piezoelectric elements, which enable the fine movements necessary for bringing the probe tip in contact with the specimen. The tips of the needles are sharpened, as shown in (c), by using an FIB system.

As described later in detail, the probes can be used for a variety of applications, such as applying a large electric field and/or electric current [60, 61] and producing a large magnetic field with a magnetic needle made of a permanent magnet. The holder was also modified to carry a laser irradiation port in its left arm (arm 1), as shown in (d). A light convergence function can be added at the top of the laser

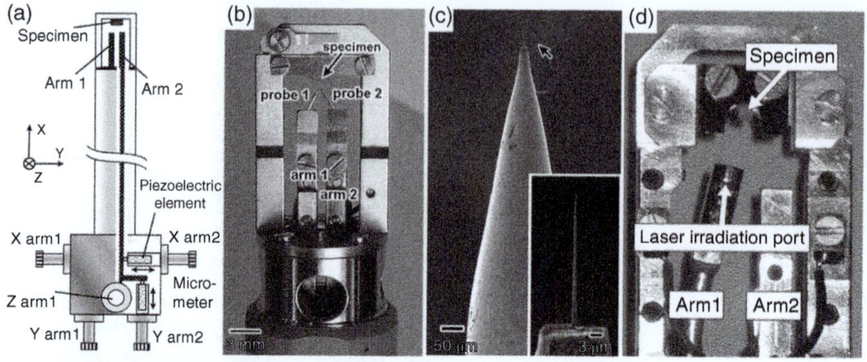

Figure 6.28 (a) Schematic representation of double-probe piezodriving holder. While specimen holder is in use, incident electrons are parallel to Z-direction. (b) Optical micrograph of top portion of specimen holder. (c) Scanning ion microscope image of Pt–Ir needle with tip sharpened with an FIB system (inset shows enlarged image of tip). (d) Top portion of modified specimen holder equipped with piezodriving probe and laser irradiation port. Source: (b–d) Shindo and Akase (2020), with permission from Elsevier.

irradiation port. A plastic fiber is placed inside arm 1 to introduce the laser beam into the holder [62].

An example application of the double-probe piezodriving holder is imaging an electric field produced by mechanical friction (e.g. triboelectricity in toner particles and photoconductors), which is an important phenomenon in electric printing technology. The charge on a toner particle has yet to be evaluated. By utilizing the present technique, the charge on toner particles with various shapes and sizes can be evaluated. Such information will be useful for producing toner particles with optimized properties. Two technical problems are encountered when using electron holography for such electric field analysis. The first is unwanted charging of the toner particles due to electron irradiation: the emission of secondary electrons makes the toner particles positively charged although mechanical friction by itself induces negative charges. To avoid this problem, a thin-plate Mo shield can be placed over the toner particles to screen them from the electron beam [63]. Although the toner particles cannot be imaged when an Mo shield is used, the charge-induced electric field outside the shield can be observed by holography, as shown in Figure 6.29b. The other problem is perturbation of the reference wave by the long-range field produced by the triboelectric charge. To prevent this electric field and free the reference wave from the perturbation, another shield (a Pt–Ir plate, connected to ground) can be placed within the hologram, as shown in (a). The position of the shields can be accurately controlled using the piezodriving arms of the specimen holder [59].

The left panel in Figure 6.30a shows a reconstructed phase image obtained using the experimental setup shown in Figure 6.29b; note that the phase shift is shown in terms of the direct phase ϕ rather than cos ϕ. Many contour lines can be seen between the toner particle of interest and the Pt–Ir shield while such lines are not evident on the right of the shield due to the successful prevention of the electric field.

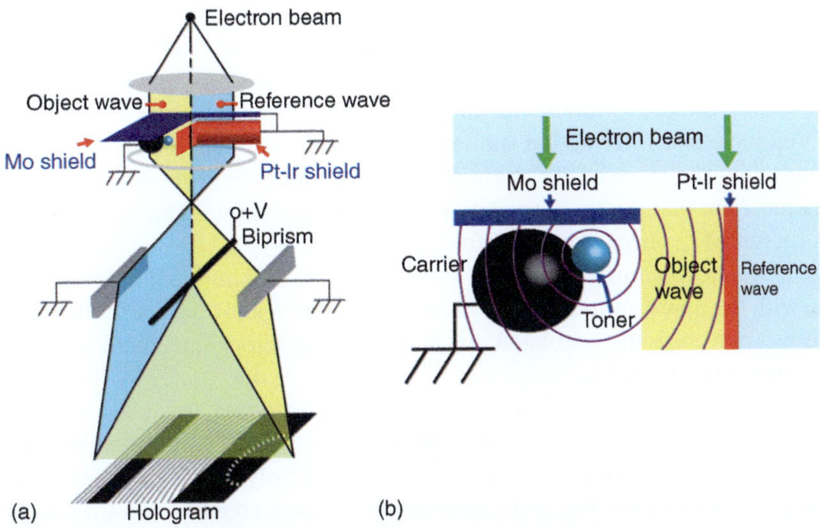

Figure 6.29 (a) Electron holography imaging process. (b) Shielding technique used to prevent electron irradiation for toner particles.

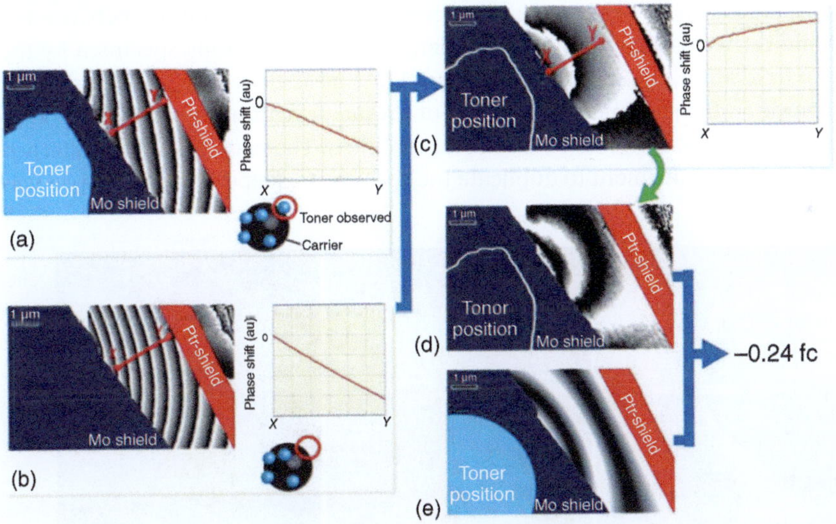

Figure 6.30 Reconstructed phase images showing triboelectricity of toner particle (a) with particle (blue region) beneath Mo shield and (b) after removing particle. (c) Result of subtracting phase information of (b) from that of (a). (d) Result of (c) presented in terms of cos ϕ. (e) Result obtained by computer simulation. Source: Shindo and Akase [1], with permission from Elsevier.

Unfortunately, as shown in Figure 6.30a, there were many positive charges under the Mo shield (refer to the negative slope shown in the right panel, which represents the phase shift measured along line X–Y). This is because the result contains contributions not only from the toner particle of interest (blue area in (a)) but also from carrier particles and/or other toner particles that are out of view.

To eliminate the undesired signal components, another reconstructed phase image, shown in Figure 6.30b, was acquired for the same field of view after removing the toner particle of interest by using part of the Pt–Ir shield. The true electric field of the toner particle (i.e. the removed toner particle) was obtained by subtracting the phase information shown in (b) from that shown in (a). The result is shown in (c). The slope of the phase shift in (c) is opposite those in (a) and (b). We thus conclude that the toner particle was negatively charged. The result shown in (c) is presented in terms of cos φ in (d).

Figure 6.30e shows simulation results for the equipotential lines around the toner particle, with the phase shift expressed in terms of cos ϕ. When an electric charge of −0.24 fC was assumed, the simulated image was in satisfactory agreement with the observed image shown in (d).

Split-illumination electron holography (explained in Section 6.1) can be used with a mask plate to analyze the electrostatic interactions between a toner particle and a carrier particle [64]. Model toner specimens containing polystyrene particles, positive or negative toner particles, and a carrier particle were attached to a Cu needle for electron microscopic observation. The determination of "positive" or "negative" was determined by the polarity of the triboelectric charges.

Figure 6.31a shows the optical system used for analyzing the electrostatic potential distributions (yellow) around a toner particle and the distributions between the toner and carrier particles. To prevent charging of the insulating specimen by the electron irradiation, the specimen was placed in a mask plate shadow. The shadow was formed on the specimen plane without Fresnel fringes by adjusting the focusing lens of the mask between the mask and specimen. The mask plate was patterned by using an FIB instrument to duplicate the specimen's shape. The mask plate was

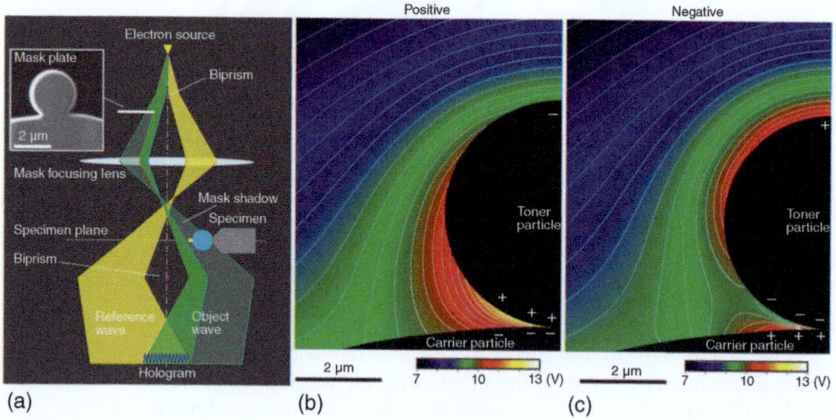

Figure 6.31 (a) Schematic of optical system for observing and analyzing electrostatic potential distributions of charged insulating specimen. Specimen was placed in shadow of mask plate to prevent charging by electron irradiation. Split-illumination electron holography was used to remove perturbation of reference wave caused by electric field from charged specimen. (b and c) Analyzed electrostatic potential distributions around (b) positive and (c) negative toner particles at specimen plane. Each contour line corresponds to potential difference of 0.2 V.

mechanically rotated and moved horizontally. Because the mask was placed in the illumination system, the effects of the secondary electrons emitted from the mask were eliminated through the narrow electron paths in the lens system between the mask and specimen positions.

Split-illumination electron holography was used to place the reference wave far from the charged specimen to minimize reference-wave perturbations due to specimen charging. The smooth modulations in the reference wave caused by long-range electric fields were simulated in the phase analysis. In the electrostatic potential analysis, a phase image reconstructed from the hologram was compared with the simulated phase image by considering various potential distributions in the specimen. Phase simulations were performed in three dimensions to calculate the phase shifts in the object wave and the reference wave. In these simulations, the electrostatic potential had axial symmetry along the axis passing through the center of the toner and carrier particles (for details, see the report [64]).

The analyzed electrostatic potential distributions around the toner particle attached to the carrier particle (shown in Figure 6.31b,c) show that the distributions were nonuniform for both the positive and negative specimens. They also indicate that the local charges on the toner particle surface had opposite values for the positive and negative specimens. The charges on the carrier surface facing the toner particle were opposite those on the toner particle surface for both specimen types. The charges on the toner particle surface far from the carrier particle were opposite those on the carrier side of the toner particle.

These findings suggest that dielectrically polarized charges contribute to the electrostatic interactions between toner and carrier particles. The detailed charge distribution on both a toner particle and a carrier particle has yet to be analyzed. The detailed mechanism of the adhesion of charged particles can be understood on the basis of quantitative information about the distribution of charges on toner particles and a carrier particle obtained using this technique, resulting in improved electrophotographic printing.

As clarified in these studies, the amount and distribution of electric charges can be evaluated without irradiating the toner and carrier specimens with incident electrons. Here, we should recall the importance of "field" by citing the words of Einstein and Infeld given in the introduction: "...it is not the charges nor particles but the field in the space between the charges and particles which is essential...."

6.7 Specimen Preparation Techniques

Among the various specimen thinning methods for TEM observation, the FIB method and ultramicrotomy are mainly explained here with examples of thin films prepared using these methods. Other methods are briefly explained.

The FIB method is one of the most widely used methods for preparing TEM specimens. This method was originally developed for repairing semiconductor

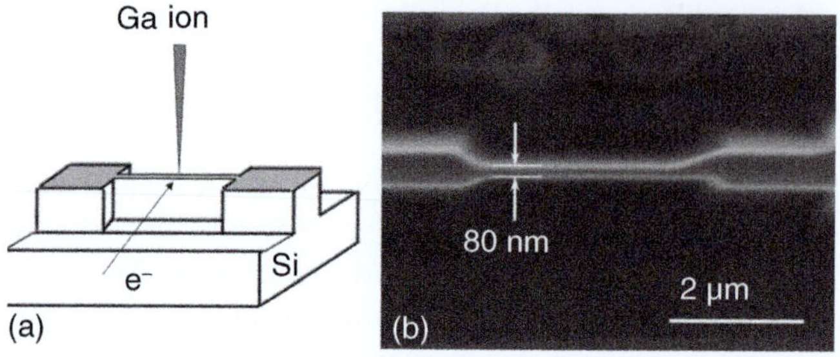

Figure 6.32 (a) Geometric configuration of incident Ga-ion beam and Co–CoO tape specimen. (b) Co–CoO thin film prepared using FIB method. Source: (b) Shindo and Akase [1], with permission from Elsevier.

devices. Ion beams are sharply focused on a small area, and the specimen is rapidly thinned by sputtering. Usually, Ga ions are used with an accelerating voltage of about 30 kV and a current of about $10\,\mathrm{A\,cm^{-2}}$. The probe size is several tens of nanometers.

Figure 6.32a illustrates the geometric configuration of the incident ion beam for thinning and the incident electron beam for TEM observation. The FIB method is currently attracting much attention as it may be especially useful for specimens that contain a boundary between different materials, which can make it difficult to thin the boundary region homogeneously by other methods, such as ion milling. By detecting the secondary electrons emitted from the specimen while irradiating it with the ion beams, a secondary-electron image of the surface can be displayed as a scanning electron microscopy image. Thus, by observing the secondary-electron image, the appropriate region for thinning can be accurately selected. Special care is required to avoid irradiation damage due to the strong ion beams and to avoid implanting of Ga ions.

Figure 6.33 shows a cross section of obliquely evaporated Co–CoO magnetic tape prepared using the FIB method (see Figure 6.32b) [65]. The upper film layers of the tape were fixed on a Si substrate to enable easy handling during the thinning process. The detailed magnetic flux distribution both outside and inside the specimen can be visualized by observing holograms and removing the electric field information due to the inner potential, as explained in Section 4.3.4 and shown in Figure 6.33a,b, respectively.

In contrast, ultramicrotomy is used mainly to prepare thin sections of soft materials such as biological specimens. Specimens of thin films or powders are usually fixed in a resin and trimmed with a glass knife before being sliced with a diamond knife. Figure 6.34 illustrates the geometric configuration of ultramicrotomy. Each time the arm holding the specimen moves up and down, it advances; in this way, the diamond knife slices the specimen, which floats on the water in a metal boat. The sliced sections drop onto the water and are retrieved using a thin wooden stick with an eyelash and placed on a special grid covered with a collodion or carbon thin film.

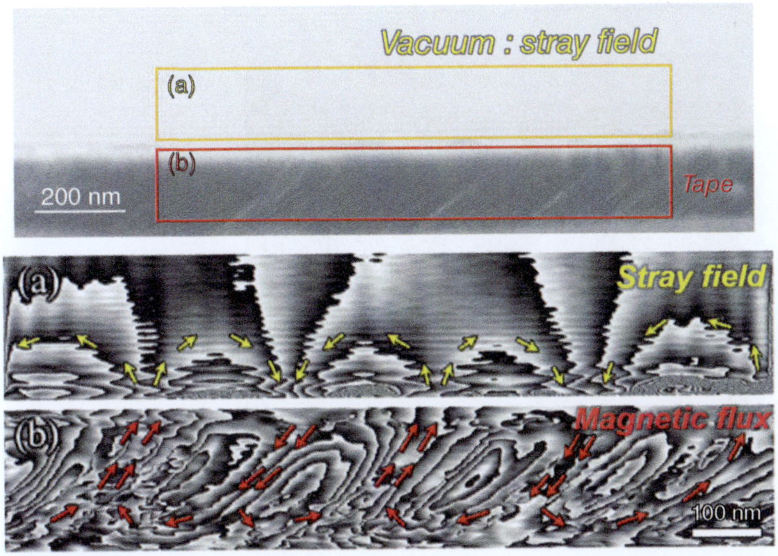

Figure 6.33 Hologram (above) and reconstructed phase images (below) of Co–CoO tape specimen. Reconstructed phase images clarify magnetic flux distribution (a) outside and (b) inside specimen. Source: Shindo and Akase [1], with permission from Elsevier.

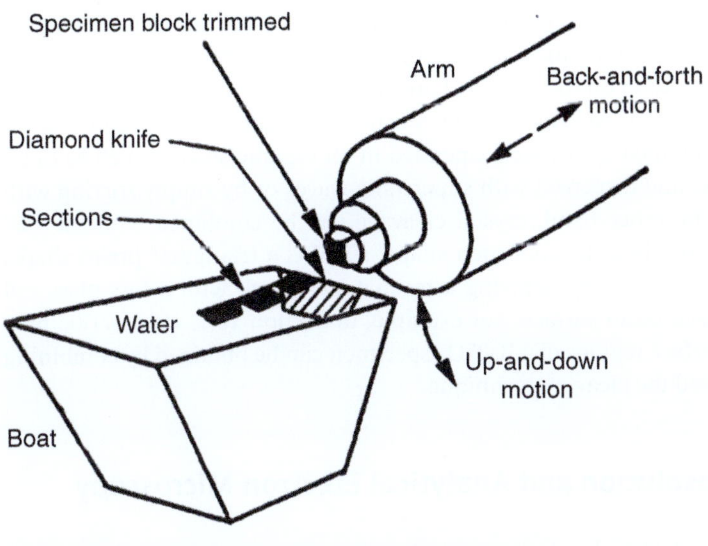

Figure 6.34 Geometric configuration of ultramicrotomy.

One of the advantages of ultramicrotomy is that a clean surface can be obtained without ion irradiation damage, as discussed in Section 10.2. For comparison with the results of the FIB method (Figure 6.32), a Co–CoO tape prepared by ultramicrotomy is shown in Figure 6.35. Although a clean surface without ion implantation can be obtained, lattice strain is frequently introduced into the sections during slicing.

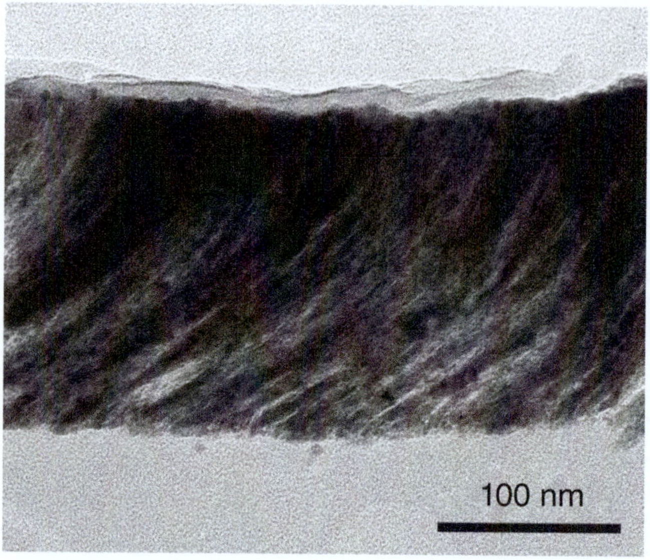

Figure 6.35 Cross-sectional view of Co–CoO magnetic tape prepared by ultramicrotomy. Source: Shindo and Akase [1], with permission from Elsevier.

A specimen with a clean surface can be obtained by applying the crushing method to oxides, ceramics, and so on. Even though this method is the simplest among the specimen preparation techniques, it can obtain thin regions (a few nanometers) with little contamination on the surface. Unfortunately, its application is limited to materials that tend to cleave. A specimen is usually crushed with an agate mortar and pestle. The obtained flakes are suspended in an organic solvent such as butyl alcohol or acetone and dispersed with supersonic waves or by simply stirring with a glass rod. On the other hand, crystal cleavage can be combined with the FIB method. After controlling the specimen shape, such as a triangular prism shape, a thin part can be cleaved by applying a mechanical shock with a tiny glass rod to obtain a localized clean surface. For example, in Section 10.1, we describe how localized clean surface regions of a $BaTiO_3$ specimen can be obtained by combining the FIB method and the cleaving technique.

6.8 High-Resolution and Analytical Electron Microscopy

To understand the electrical and magnetic properties of various advanced materials, in addition to electromagnetic field visualization by electron holography and Lorentz microscopy, characterization of the microstructures of such materials is fundamentally important. The high-resolution electron microscopes currently in use have a resolution of less than 0.1 nm. Using microprobes with a size of less than 0.1 nm and beam scanning systems enables STEM to be used for elemental mapping at the atomic level. The basic principles and their applications are as follows.

6.8.1 Conventional Microscopy and High-Resolution Electron Microscopy

To investigate an electron microscope image, we first observe the electron diffraction pattern. Then, by passing the transmitted beam or one of the diffracted beams through the objective aperture and switching the microscope to imaging mode, we can observe the image with enhanced contrast, enabling precipitates and lattice defects to be easily identified using conventional methods.

As shown in Figure 6.36a, the observation mode using a transmitted beam is called the “bright-field method,” and the image observed is called a “bright-field image.” When one diffracted beam is selected (b), it is called the “dark-field method,” and the observed image is called a “dark-field image.” The contrast in these images is attributed to the change in the amplitude of the transmitted beam or the diffracted beam due to absorption and scattering in the specimen. Thus, the image contrast is called the “absorption–diffraction contrast” or “amplitude contrast.”

It is also possible to form electron microscope images by selecting more than two beams on the back focal plane by using a large objective aperture, as shown in Figure 6.36c. This observation mode is called “high-resolution electron microscopy,” and the image observed is called a “high-resolution electron microscope image.” Since the contrast in high-resolution microscope images is due to the difference in phase between the transmitted and diffracted beams, the contrast is called the “phase contrast.”

Figure 6.37a,c shows high-resolution images of $Co_{71.5}Zr_{9.2}O_{19.3}$, $Co_{59.9}Zr_{10.3}O_{29.8}$, and $Co_{52.9}Zr_{12.0}O_{35.1}$ thin films, respectively [66]. It is seen that all of these films have a nanogranular structure, and Co grains indicated by arrows are distributed in the amorphous matrix. From Figure 3, it can be pointed out that there is no definite crystal orientation relationship among the Co grains. It infers that the magnetic anisotropy of these nanogranular thin films is caused by an induced magnetic anisotropy that is not originated from a preferred crystalline orientation. These high-resolution images clarify the difference of sized distribution of Co grains. Combining these results with domain structures obtained by electron holography and Lorentz microscopy, their magnetic properties were discussed [66] (Figure 6.37).

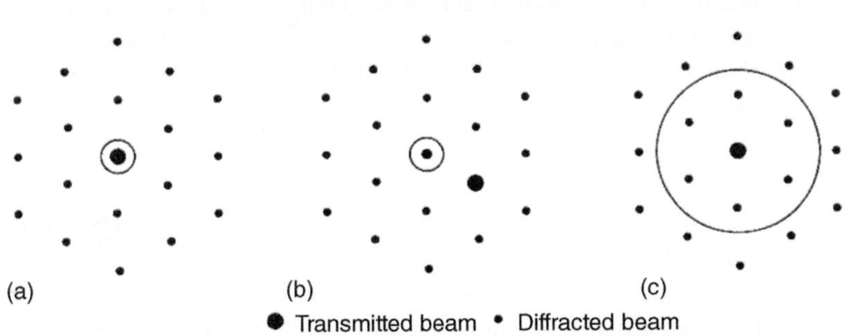

Figure 6.36 Three observation modes in electron microscopy using an objective aperture. Center of objective aperture is assumed to be set to optical axis. (a) Bright-field method. (b) Dark-field method. (c) High-resolution electron microscopy.

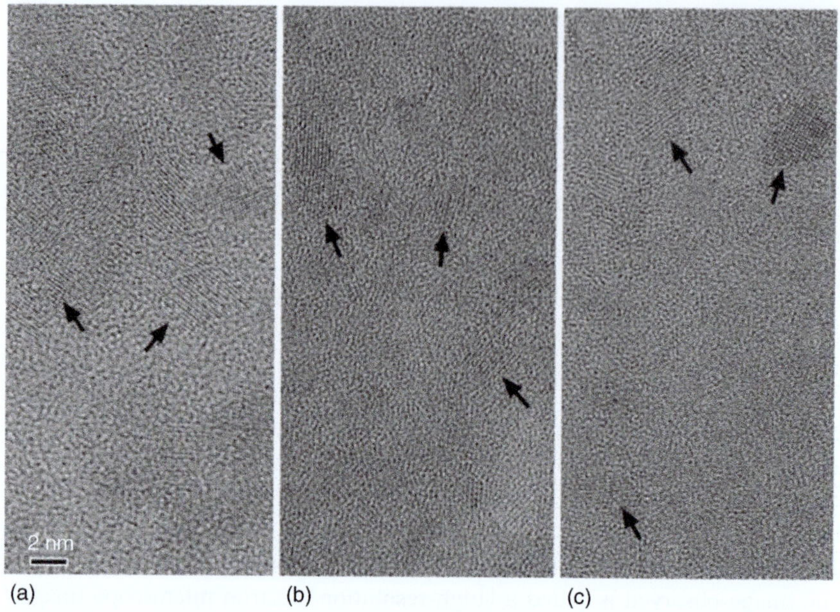

Figure 6.37 High-resolution images of $Co_{71.5}Zr_{9.2}O_{19.3}$ (a), $Co_{59.9}Zr_{10.3}O_{29.8}$ (b), and $Co_{52.9}Zr_{12.0}O_{35.1}$ (c) thin films. Liu et al. [66], with permission from Elsevier.

6.8.2 High-Angle Annular Dark-Field Method

In addition to these techniques, high-angle annular dark-field (HAADF)–STEM is widely utilized. Figure 6.38 shows the principle of HAADF–STEM. According to Pennycook [67], the partial scattering cross section of the electrons distributed in the annular shaded area can be obtained by integrating the Rutherford scattering intensity from scattering angle θ_1 to θ_2 using

$$\sigma_{\theta 1,\theta 2} = \left(\frac{m'}{m}\right)^2 \frac{Z^2\lambda^4}{4\pi^3 a_0^2}\left(\frac{1}{\theta_1^2+\theta_0^2} - \frac{1}{\theta_2^2+\theta_0^2}\right) \tag{6.9}$$

where m' is the electron mass, λ is the electron wavelength, m is the electron rest mass, a_0 is the Bohr radius, Z is the atomic number, and θ_0 is the Born scattering angle. If the number of atoms in a unit volume of the specimen is N, scattering intensity I_s given by

$$I_s = \sigma_{\theta 1,\theta 2} NtI \tag{6.10}$$

where t and I indicate the specimen thickness and the incident electron intensity, respectively. Looking at Eqs. (6.9) and (6.10), we can see that the signal intensity of an HAADF–STEM image is proportional to the square of the atomic number Z. Therefore, the image contrast strongly depends on Z, and a HAADF–STEM image is sometimes called a "Z contrast image" or "Z^2 contrast image."

Figure 6.39a shows a HAADF-STEM image of an $Sm_2(Fe_{0.95}, Mn_{0.005})_{17}N_{4.2}$ hard magnetic material [68]. The specimen consists of amorphous and crystal regions.

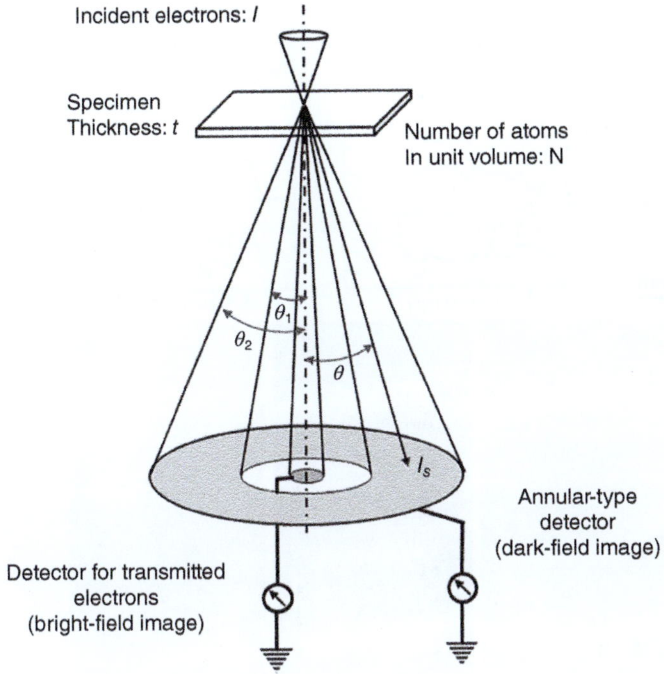

Figure 6.38 Principle of HAADF microscopy.

The amorphous regions appear dark, indicating that they consist of relatively light elements, as clarified below. Figure 6.39b,c is described in the next section.

6.8.3 Analytical Electron Microscopy

Electron energy-loss spectrometers and energy-dispersive X-ray spectrometers, typical analytical instruments installed in transmission electron microscopes, are utilized extensively. Here, two standard analytical methods, i.e. electron energy-loss spectroscopy (EELS) and energy-dispersive X-ray spectroscopy (EDS), are presented.

The principles of EELS and EDS can be explained using an inelastic electron scattering process, i.e. excitation of an inner-shell electron. Figure 6.40a shows the change in electronic structure due to excitation of an electron in the K shell. The resultant energy-loss spectrum and X-ray spectrum observed are shown in (b) and (c), respectively.

Here we consider the case in which an incident electron gives energy to the specimen, and the electron in the K shell (1s orbital) is excited. Because the energy levels below the Fermi level are all occupied by electrons in the ground state, one of the electrons in the K shell can transit to the unoccupied state above the Fermi energy. Thus, when the electron loses more energy than ΔE, which corresponds to the energy difference between the energy at the K shell and the Fermi energy, the probability of a transition from the K shell to the unoccupied density of states increases drastically, and eventually a sharp peak appears at energy ΔE in the

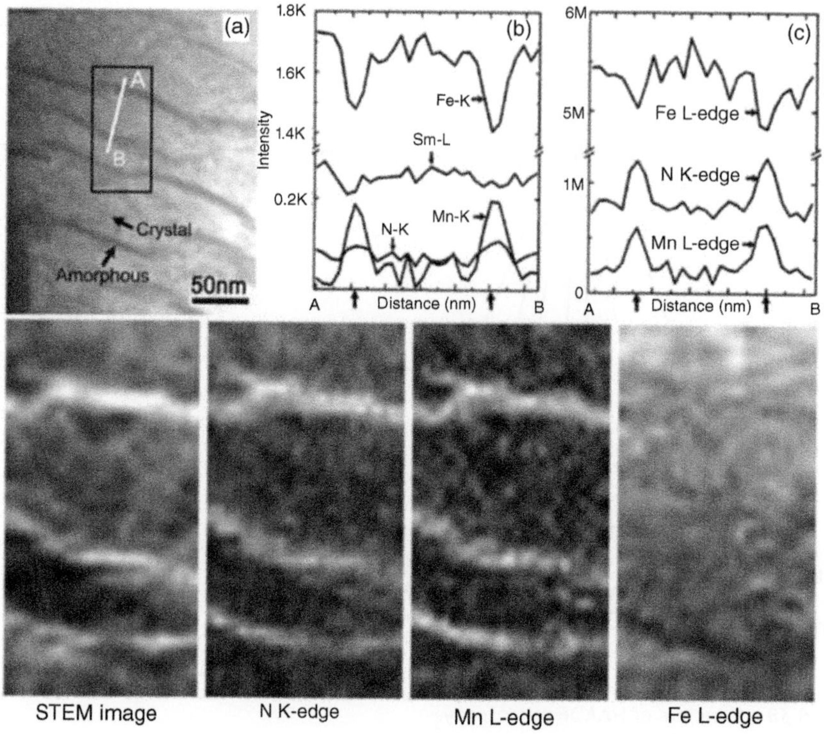

Figure 6.39 (a) HAADF–STEM image of $Sm_2(Fe_{0.95}, Mn_{0.05})_{17}N_{4.2}$. (b and c) One-dimensional intensity profiles of line A–B in (a) by EDS. Lower panels show STEM image and two-dimensional elemental mapping images of elemental distribution (N, Mn, Fe) obtained from rectangular area in (a). Source: Yasuhara et al. [68], with permission from Elsevier.

energy-loss spectrum as shown in Figure 6.40b. In this excitation of the inner-shell electron, the peak tends to accompany the tail in a higher energy region. From this shape, the peak appearing in the energy-loss spectrum is generally called an "edge."

Because the threshold energy of the edge is specific to each material, the specimen can be identified by its energy value, ΔE. Furthermore, information about the content of the element being observed can be obtained from the integrated intensity of the edge. Moreover, from an accurate value of the threshold energy and the shape of the edge, information about the chemical bond can be obtained. There are also other atom excitation processes due to incident electrons, such as interband transition and collective excitation of valence electrons (plasmon excitation).

When an atom changes from the excited state to the ground state, the surplus energy is emitted as a characteristic X-ray or an Auger electron. In both cases, one of the electrons in a higher energy level transits to the hole in the lower-energy level in a manner satisfying the selection rule. Like EELS, the energy of the characteristic X-ray can be used to specify a constituent element, as the energy at an X-ray peak position is specific to an element. Additionally, the composition of the material can be evaluated from the integrated intensity. The characteristic X-ray emission

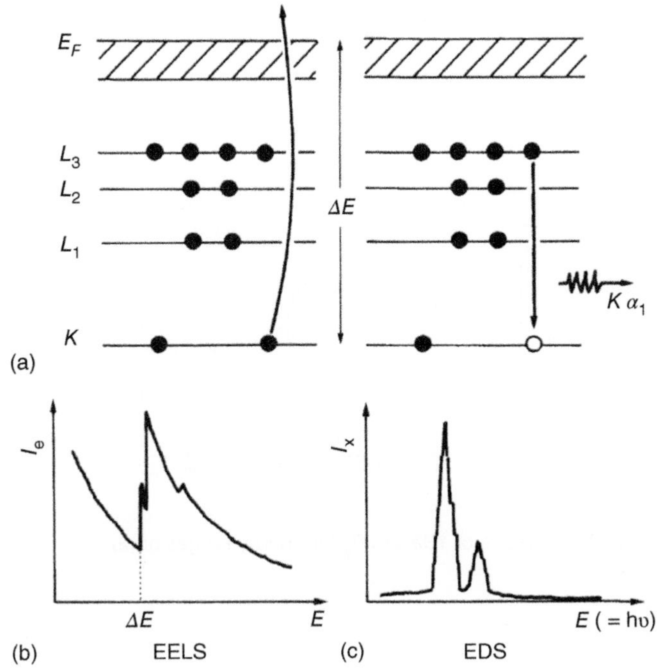

Figure 6.40 (a) Inner-shell electron excitation. (b) Resultant electron energy-loss spectrum. (c) Energy dispersive X-ray spectrum.

resulting from transition of an electron from the L_3 shell to the K shell is illustrated in Figure 6.40c, and the X-ray emitted is called a "characteristic $K_{\alpha 1}$ X-ray." Several other characteristic X-rays are also frequently used for compositional analysis, such as $K_{\beta 1}$ and $L_{\alpha 1}$, which correspond to the transition from the M_3 shell to the K shell and from the M_5 shell to the L_3 shell, respectively.

As explained above, EELS and EDS are related to the same excitation process (i.e. inner-shell electron excitation), so it might be considered that similar information can be obtained from them. However, owing to the large differences in the background height and the resolution of the spectra, the information obtained differs. For example, the resolution of EELS is around 1 eV whereas that of EDS is around 150 eV. To illustrate the situation, an electron energy-loss spectrum [15] and a characteristic X-ray spectrum obtained from YBa_2Cu_3O superconductor are presented in Figures 6.41 and 6.42, respectively.

Although the electron energy-loss spectrum can generally be observed in the energy range from 0 to about 2 keV in a conventional EELS system, only a limited energy region of 500–950 eV is presented in Figure 6.41a. Small peaks of an oxygen K-edge, a Ba-$M_{4,5}$ edge, and a Cu-$L_{2,3}$ edge are evident on a large background. In contrast, in (b), three energy-loss spectra at the oxygen K-edge observed at different temperatures are presented in the energy range 520–570 eV. Whereas the oxygen content decreases with an increase in temperature, resulting in decreased hole content, the peak at 528 eV of the energy-loss spectrum decreases with a decrease

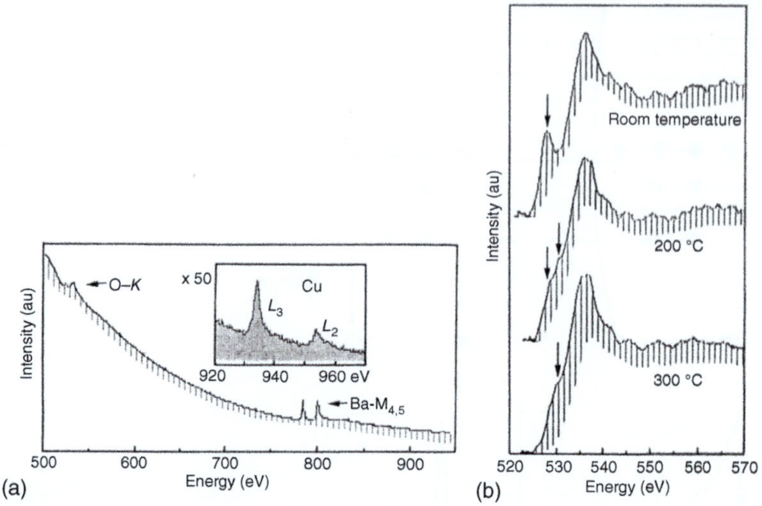

Figure 6.41 Electron energy-loss spectra of $YBa_2Cu_3O_y$ for energy ranges of (a) 500–950 eV and (b) 520–570 eV.

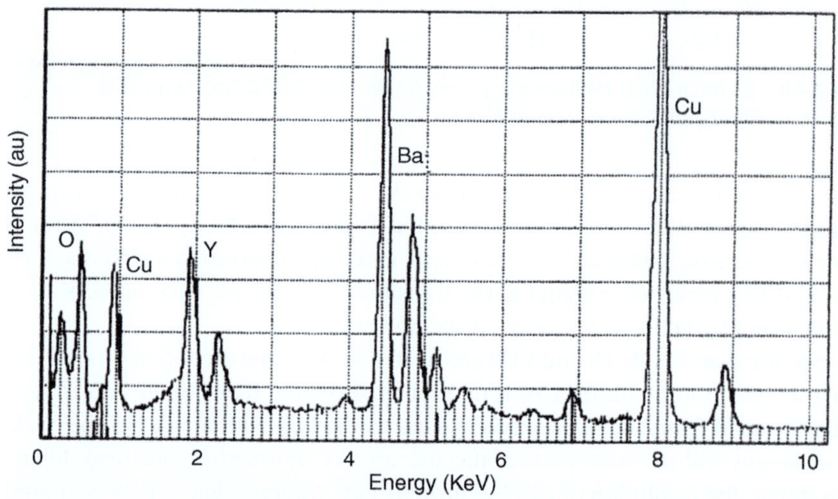

Figure 6.42 Characteristic X-ray spectrum of $YBa_2Cu_3O_y$.

in hole content. Thus, EELS shows not only compositional information but also information on the electronic state, especially the unoccupied density of states. In contrast, as shown in Figure 6.42, based on the characteristic X-ray spectrum in the energy range 0–10 keV (a spectrum available up to 20 keV), quantitative compositional information can be easily obtained owing to the low background of the spectrum.

EDS and EELS can both be used for elemental mapping, as shown by the intensity profiles of one-dimensional elemental mapping images shown in Figure 6.39b,c, respectively, corresponding to line A–B in (a). In the EDS intensity profiles, the amorphous regions are rich in Mn, indicated by the black arrows between "A" and "B" and poor in Fe. In the EELS intensity profiles, the amorphous regions are rich in N and Mn. These features are clearly visualized in the two-dimensional elemental mapping images with EELS shown at the bottom of Figure 6.39, in which the amorphous phases enriched with N and Mn appear as brighter regions. The large coercivity of this magnetic material is considered to result from domain wall pinning in the amorphous boundary regions.

Next, we demonstrate the elemental mapping for two Sm–Co permanent magnets to clarify the distribution of the additives. It is well known that the properties of Sm–Co magnets are very sensitive to the amounts of additives and the heat treatment. To analyze the distribution, EDS elemental mapping was carried out with a 0.7-nm probe. The images at the top left in Figure 6.43a,b are STEM images of a magnet that was isothermal aged for two hours with a coercive force of 40 kA m^{-1} and of one that was step aged with 836 kA m^{-1}, respectively. Characteristic X-ray images of Cu, Fe, and Zr are presented in both figures. As evident in the figures, both magnets contain small amounts of Cu, Fe, and Zr and consist of matrix phases of 2 : 17 R, 1 : 5 H, precipitate phases, and the z phase. The additives are clearly segregated by the step aging. In particular, Zr is clearly located at the *z* phase with a thickness of about 1 nm, and Cu is segregated at the cross point as indicated by an arrow in domains of the 1 : 5 H phase in (b). These results indicate that step-aging treatment leads to chemical partitioning among the phases without any microstructural changes, resulting in high coercive force [27, 69].

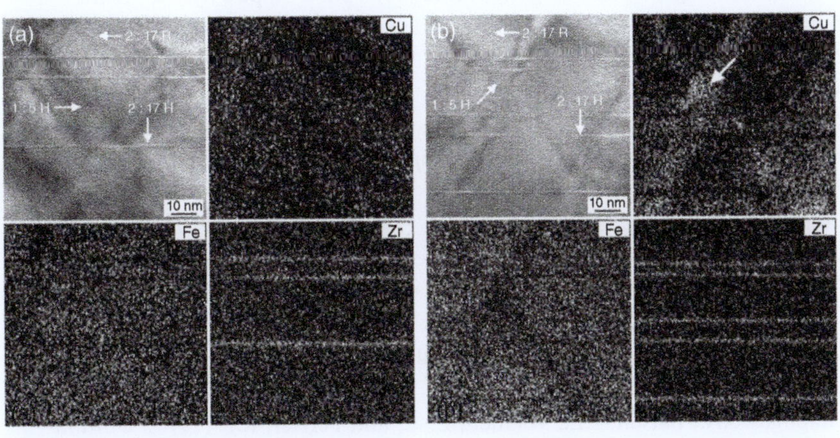

Figure 6.43 Elemental mapping images of Sm–Co magnet. (a) Isothermal aging. (b) Stepped aging. Top left image of each is STEM image.

References

1 Shindo, D. and Akase, Z. (2020). *Mater. Sci. Eng. R Reports* 142: 100564. https://doi.org/10.1016/j.mser.2020.100564.
2 Tanigaki, T., Inada, Y., Aizawa, S. et al. (2012). *Appl. Phys. Lett.* 101: 043101.
3 Harada, K., Tonomura, A., Togawa, Y. et al. (2004). *Appl. Phys. Lett.* 84: 3229.
4 Tanigaki, T., Aizawa, S., Park, H.S. et al. (2014). *Ultramicroscopy* 137: 7.
5 Kelly, P.M., Jostsons, A., Blake, R.G., and Napier, J.G. (1975). *Phys. Status Solidi A* 31: 771.
6 Allen, S.M. (1981). *Philos. Mag. A* 43: 325.
7 Tanaka, M. and Terauchi, M. (1985). *Convergent-Beam Electron Diffraction.* Tokyo: JEOL LTD.
8 Spence, J.C.H. and Zuo, J.M. (1992). *Electron Microdiffraction*. New York: Plenum Press.
9 Nishino, D., Nakafuji, A., Yang, J.-M., and Shindo, D. (1998). *ISIJ Int.* 38: 1369.
10 Akase, Z., Shindo, D., Inoue, M., and Taniyama, A. (2007). *Mater. Trans.* 48: 2626.
11 Inada, Y., Akase, Z., Shindo, D., and Taniyama, A. (2012). *Mater. Trans.* 53: 1330.
12 Hÿtch, M., Houdellier, F., Hüe, F., and Snoeck, E. (2008). *Nature Lett* 453: 1086.
13 Murakami, Y., Niitsu, K., Tanigaki, T. et al. (2014). *Nat. Commun.* 5: 1.
14 Shindo, D. and Akase, Z. (2020). *Mater. Sci. Eng., R* 142: 100564.
15 Shindo, D. and Oikawa, T. (2002). *Analytical Electron Microscopy for Materials Science.* Tokyo: Springer-Verlag.
16 Kappert, H. (1970). *Z. Angew. Phys.* 29: 139.
17 Kappert, H. and Schmiesing, P. (1971). *Phys. Status Solidi A* 4: 737.
18 Suzuki, T. and Hubert, A. (1970). *Phys. Status Solidi* 38: K5.
19 Suzuki, T., Hiraga, K., and Sagawa, M. (1984). *Jpn. J. Appl. Phys.* 23: L421.
20 Yang, J.M., Shindo, D., Lim, S.H. et al. (1998). *Electron Microsc., ICEM* 14: 559.
21 Perkins, R.S., Gaiffi, S., and Menth, A. (1975). *IEEE Trans. Magn.* MAG-11: 1431.
22 Nagel, H. (1979). *J. Appl. Phys.* 50: 1026.
23 Kittel, C. (1949). *Rev. Mod. Phys.* 21: 541.
24 Fuller, H.W. and Hale, M.E. (1960). *J. Appl. Phys.* 31: 238.
25 Wade, R.H. (1966). *J. Appl. Phys.* 37: 366.
26 Yang, J.M., Shindo, D., and Hiroyoshi, H. (1997). *Mater. Trans., JIM* 38: 363.
27 Mishra, R.K., Thomas, G., Yoneyama, T. et al. (1981). *J. Appl. Phys.* 52: 2517.
28 Chapman, J.N., Johnston, A.B., Heyderman, L.J. et al. (1994). *IEEE Trans. Magn.* 30: 4479.
29 McVitie, S., Chapman, J.N., Zhou, L. et al. (1995). *J. Magn. Magn. Mater.* 148: 232.
30 Chapman, J.N., Batson, P.E., Waddell, E.M., and Ferrier, R.P. (1978). *Ultramicroscopy* 3: 203.
31 Tsuno, K. (1988). *Rev. Solid State Mater. Sci.* 2: 623.
32 Morrison, G.R., Chapman, J.N., Craven, A.J. (1980). Electron Microscopy and Analysis 1979. *Proceedings of EMAG79. Brighton, UK 3–6 September 1979 EMAG 79*, CRC, Taylor&Francis Group. 257.

33 Tsuno, K., Inoue, M., and Ueno, K. (1989). *Mater. Sci. Eng., B* 3: 403.
34 Tsuno, K. and Inoue, M. (1984). *Optik* 67: 363.
35 Shibata, N., Kohno, Y., Nakamura, A. et al. (2019). *Nat. Commun.* 10: 2308.
36 Wu, M., Tafel, A., Hommelhoff, P., and Spiecker, E. (2019). *Appl. Phys. Lett.* 114: 013101.
37 Teague, M.R. (1983). *J. Opt. Soc. Am.* 73: 1434.
38 Van Dyck, D. and Coene, W. (1987). *Optik* 77: 125.
39 Bajt, S., Barty, A., Nugent, K.A. et al. (2000). *Ultramicroscopy* 83: 67.
40 De Graef, M. and Zhu, Y. (2001). *J. Appl. Phys.* 89: 7177.
41 Volkov, V.V. and Zhu, Y. (2004). *Ultramicroscopy* 98: 271.
42 McVitie, S. and Cushley, M. (2006). *Ultramicroscopy* 106: 423.
43 Paganin, D. and Nugent, K.A. (1998). *Phys. Rev. Lett.* 80: 2586.
44 Ishizuka, K. and Allman, B. (2005). *J. Electron Microsc.* 54: 191.
45 HREM Research Inc., Tokyo, Japan. www.hremresearch.com.
46 Shindo, D., Park, Y.-G., Liu, Y. et al. (ed.) (2006). *Handbook of Advanced Magnetic Materials*, 24–65. New York: Springer Science + Business Media, Inc.
47 Tonomura, A. (1999). *Electron Holography*, 2e. Springer-Verlag.
48 Huber, E.E., Smith, D.O., and Goodenough, J.B. (1958). *J. Appl. Phys.* 29: 294.
49 Ramstöck, K., Hartung, W., and Hubert, A. (1996). *Phys. Status Solidi A* 155: 505.
50 Rave, W. and Hubert, A. (1998). *J. Magn. Magn. Mater.* 184: 179.
51 Murakami, Y., Tanigaki, T., Sasaki, T.T. et al. (2014). *Acta Mater.* 71: 370.
52 Park, H.S., Hirata, K., Yanagisawa, K. et al. (2012). *Small* 8: 3640.
53 Xia, W.X., Kim, J.J., Yogo, T. et al. (2008). *J. Magn. Magn. Mater.* 320: 3011.
54 Gao, Y., Shindo, D., Bao, Y., and Krishnan, K. (2007). *Appl. Phys. Lett.* 90: 233105.
55 Takeno, Y., Murakami, Y., Sato, T. et al. (2014). *Appl. Phys. Lett.* 105: 183102.
56 Mathur, N., Stolt, M.J., Niitsu, K. et al. (2019). *ACS Nano* 13: 7833.
57 Tanigaki, T., Akashi, T., Sugawara, A. et al. (2017). *Sci. Rep.* 7: 16598.
58 Haider, M., Hartel, P., Mller, H. et al. (2010). *Microsc. Microanal.* 16: 393.
59 Murakami, Y., Kawamoto, N., Shindo, D. et al. (2006). *Appl. Phys. Lett.* 88: 223103.
60 Murakami, Y., Yano, T., Shindo, D. et al. (2007). *Metall. Mater. Trans. A* 38: 815.
61 Kawamoto, N., Murakami, Y., and Shindo, D. (2010). *J. Appl. Phys.* 107: 044309.
62 Shindo, D., Takahashi, K., Murakami, Y. et al. (2009). *J. Electron Microsc.* 58: 245.
63 Okada, H., Shindo, D., Kim, J.J. et al. (2007). *J. Appl. Phys.* 102: 054908.
64 Tanigaki, T., Sato, K., Akase, Z. et al. (2014). *Phys. Lett.* 104: 131601.
65 Xia, W.X., Tohara, K., Murakami, Y. et al. (2006). *IEEE Trans. Magn.* 42: 3252.
66 Liu, Z., Shindo, D., Ohnuma, S., and Fujimori, H. (2003). *J. Magn. Magn. Mater.* 262: 308.
67 Pennycook, S.J., Berger, S.D., and Culbertson, R.J. (1986). *J. Microsc.* 144: 229.
68 Yasuhara, A., Park, H.S., Shindo, D. et al. (2005). *J. Magn. Magn. Mater.* 295: 1.
69 Shindo, D., Park, Y.G., Aoyama, Y. (2001). 8th Annual International Conference on Composites Engineering (5–11 August 2001), Tenerife, Canary Islands.

Part III

Application

7

Electric Field Analysis

7.1 Measurement of Inner Potential

One of the most basic applications of electron holography is evaluating the mean inner potential. From Eq. (2.2), when a specimen is chemically homogeneous and nonmagnetic, the phase shift of the electron is related to the specimen thickness t:

$$\phi(x, y) = \sigma\varphi(x, y)t \tag{7.1}$$

This equation indicates that the mean inner potential can be directly determined from a measurement of the phase shift when the thickness and interaction contrast are known. The specimen thickness can be determined by cross-sectional observation, electron energy-loss spectroscopy (EELS) [1–4], or convergent-beam electron diffraction [5–9], among other methods.

7.1.1 Diamond-Like Carbon

Figure 7.1a shows a cross-sectional view of a diamond-like carbon (DLC) film deposited onto a Si substrate [11]; the film thickness can be directly determined from the cross-sectional view. Figure 7.1b shows an electron hologram of the DLC film after it was removed from the Si substrate by chemical etching; the hologram was recorded by aligning the incident beam perpendicular to the film plane. A red dotted line outlines the specimen edge. Figure 7.1c shows the phase shift measured along line X (in vacuum) and line Y (crossing both the vacuum and the specimen), as indicated in Figure 7.1b. The fringes were scanned in the direction of the white arrow in Figure 7.1b. The peak shifts of interference fringes were plotted as a function of the peak number. By analyzing the phase shift in the DLC film, we evaluated the inner potential as 10.8 ± 0.9 V. Notably, after the mean inner potential is known, Eq. (7.1) can be used to determine the thickness of a fabricated film. This method has been demonstrated to be applicable to DLC films as thin as 5 nm.

7.1.2 SiO_2 Particles

From Eq. (7.1), the thickness can be evaluated if the mean inner potential is known. The mean inner potential can be calculated from the structure factor; however,

Material Characterization using Electron Holography, First Edition. Daisuke Shindo and Takeshi Tomita.

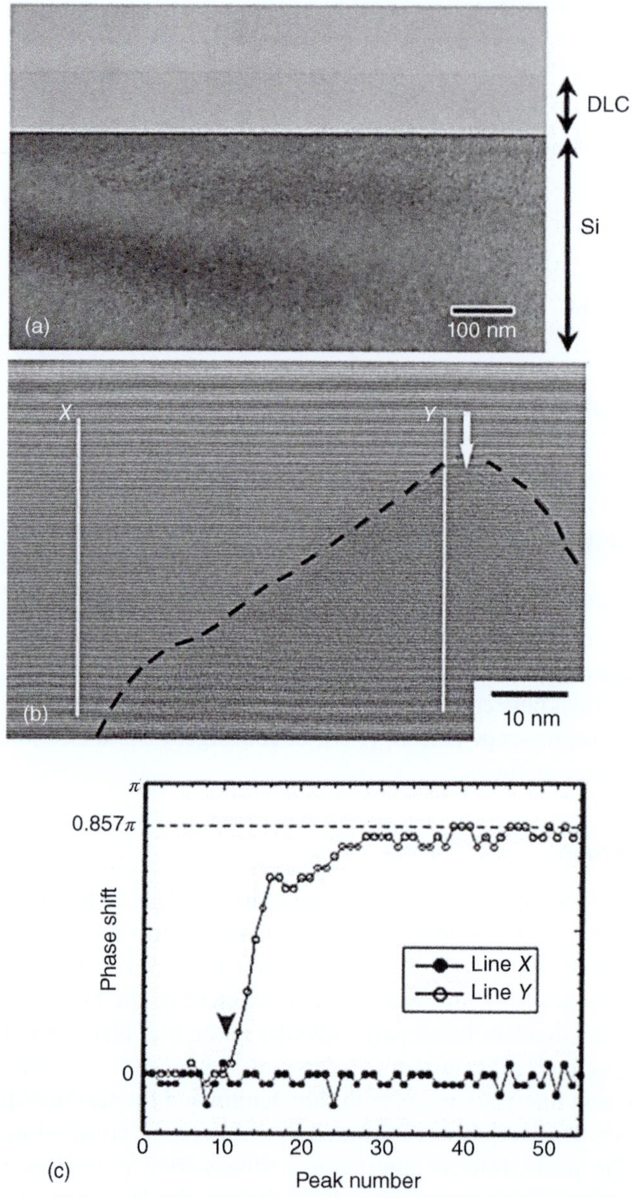

Figure 7.1 (a) Cross-sectional view of a DLC film. (b) Electron hologram of the DLC film. Dotted line outlines the specimen edge. (c) Phase shift observed along lines X and Y indicated in (b). Source: (b and c) Shindo and Akase [10], with permission from ELSEVIER.

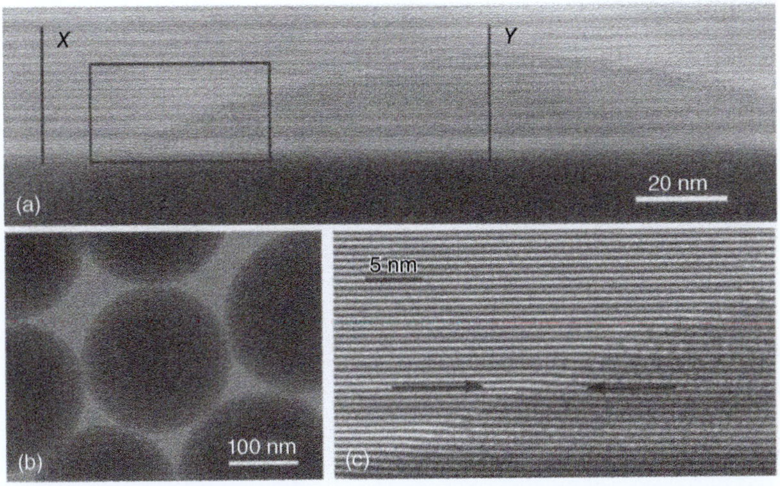

Figure 7.2 (a) Electron hologram of an amorphous SiO_2 particle. (b) Conventional TEM image of spherical SiO_2 particle. (c) Enlarged hologram of the region indicated by a rectangle in (a).

the accuracy of the structure factor directly depends on the scattering factors of the constituent elements. Notably, the mean inner potential can be determined experimentally if the shape of the specimen is known. As an example, the determination of the mean inner potential of amorphous SiO_2 is demonstrated here. Figure 7.2a shows an electron hologram of an amorphous SiO_2 particle. The interference fringes are on a part of the particle whose spherical shape is shown in a conventional TEM image (Figure 7.2b). In the enlarged hologram in Figure 7.2c, the interference fringes shift at the particle edge, where the thickness increases drastically. Figure 7.3 shows the phase shift evaluated from the interference fringes at lines X and Y. Although the phase shift at vacuum (X) is zero, that at the particle increases drastically from the particle edge. Because the particle exhibits a spherical shape, the thickness of any part of the particle can be easily estimated. Dotted lines correspond to the phase shift calculated under assumed mean inner potentials of 10.5, 11.5, and 12.5 V. From a comparison of the experimental and calculated data, the mean inner potential of amorphous SiO_2 is evaluated to be 11.5 V. With this mean inner potential, the thickness of amorphous SiO_2 can be evaluated from holograms without knowing its shape. However, the effects of charging should be carefully considered. When relatively strong electron irradiation is used for observation, the charging effect is enhanced because of the emission of secondary electrons. As shown in Figure 7.4a, the interference fringes are shifted because of the charging effect, as indicated by arrows. A simulated hologram that includes the charging effect is presented in Figure 7.4b. The charging effect and secondary electron motions are extensively studied in Part 4.

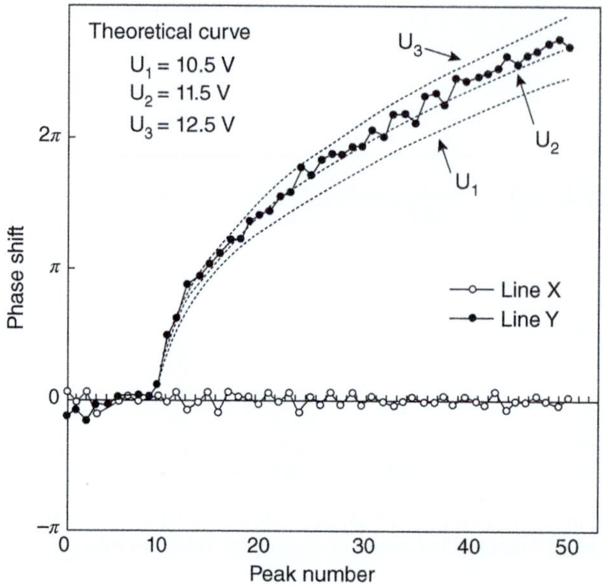

Figure 7.3 Phase shift evaluated from the interference fringes at lines *X* and *Y*.

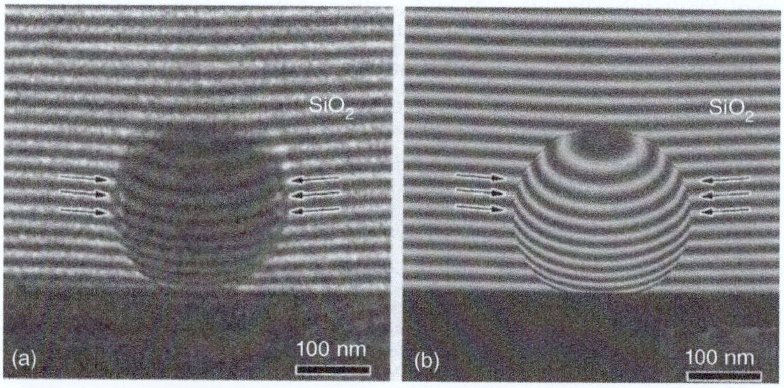

Figure 7.4 Hologram of an amorphous SiO_2 particle of 250 nm diameter (a) and its simulation (b). Arrows indicate the shift of interference fringes near the particle surface due to the charging effect.

7.1.3 p–n Junctions and Low-Dimensional Materials

We next discuss another important topic related to evaluating the mean inner potential. Following incipient studies such as those of Rau et al. [12] and Wang et al. [13], the electrostatic potential mapping of p–n junctions in semiconductor devices has attracted considerable attention. Yoo et al. [14] have also performed detailed analyses by considering the effect of the change in thickness in the field of view, a dead layer formed on the surface, and the relationship between the total thickness and the observed phase shift. Notably, three-dimensional reconstruction of the electrostatic potential in semiconductor samples has been achieved using a tomography technique [15, 16] For example, Twitchett-Harrison et al. [15] demonstrated tomographic reconstruction of the electrostatic potential near the p–n junction of a Si sample. In the case of low-dimensional materials, den Hertog et al. [17] focused on a Si nanowire and revealed the dopant profile along the growth direction by electron holography. Ichikawa [18] observed an anomalously large phase shift in nanometer-sized catalytic Au particles. Notably, accurate measurement of the inner potential requires careful consideration of a slight charging effect resulting from the electron irradiation. Recently, Wolf et al. [19] studied intrinsic electrostatic potential in three dimensions using electron holographic tomography. Ikarashi et al. [20] investigated the response of the electrostatic potential distribution within a metal–oxide–semiconductor field-effect transistor (MOSFET) to an external electric field by electron holography. Yazdi et al. [21] characterized an electrically biased Si p–n junction by measuring its electrostatic potential, electric field, and charge density distribution under working conditions using electron holography. Han et al. [22] used electron holography to measure the electrostatic potential and electric field of ferroelectric thin films to reveal the origin of non-switchable electric domains. Migunov et al. [23] measured the one-dimensional charge density distribution along an electrically biased Fe atom probe needle using a model-independent approach based on electron holography. Akase and Shindo [24] discussed the effect of dynamical diffraction on evaluating the mean inner potential. As noted in Section 6.1, for accurate analysis of electric potential distributions, split illumination with a multiple-biprism system can be effectively used.

Regarding analysis of the inner potential, quantum wells have also been studied by electron holography. For example, Cai and Ponce [25] reported a potential drop of 0.6 V at the interface between GaN and $In_{0.52}Ga_{0.48}N$ (within a quantum well of $GaN/In_{0.52}Ga_{0.48}N/GaN$). On the basis of their holographic observations, they also reported the presence of a substantial electric field of $-2.2\,MV\,cm^{-1}$ in the quantum well. The electric field was produced by both the spontaneous polarization in the wurtzite lattice (without inversion symmetry) and the piezoelectric charges produced in the strained heterostructure. For additional information on the progress in quantum-well analysis, see [26, 27].

7.2 Electric Field Analysis of Precipitates in Multilayer Ceramic Capacitor

Kawamoto et al. have demonstrated a unique characterization method based on electron holography for determining the electrical properties of submicrometer precipitates produced in multilayer ceramic capacitors (MLCCs) [28]. They showed that equipotential contour lines present essential information about the electrical conductivity of submicrometer-scale precipitates formed in commercial $BaTiO_3$ MLCCs. Researchers have reported that added Mg and rare-earth elements play an important role in controlling the temperature dependence of the dielectric constant [29]. However, the addition of such elements induces complex precipitation within the dielectric substance (e.g. a Cr-rich phase [30], Si-rich phase [31], and others). Because these precipitates have dielectric constants that differ substantially from that of the $BaTiO_3$ matrix, they are likely to degrade the performance of the capacitor. Thus, studying the local electrical properties in the region surrounding submicrometer-scale precipitates is important.

In their study, Kawamoto et al. used electron holography with handling microprobes in a transmission electron microscope [32–34] to characterize submicrometer-scale precipitates formed in MLCCs [35] and identify the objects that cause harmful electric breakdowns. Figure 7.5a shows a schematic of a thin-foil specimen (20 μm × 6 μm × 100 nm) with a stacked configuration comprising a Ni cathode, dielectric substance ($BaTiO_3$), and a Ni anode. Note that the cathode and anode are physically connected to Pt prongs, which are contacted by metallic PtIr probes when an electrical voltage is applied to the specimen. Figure 7.5b shows an ADF–STEM image of the specimen. A precipitate appearing with dark contrast is indicated by a green arrow. Energy-dispersive X-ray spectroscopy (EDS) analysis

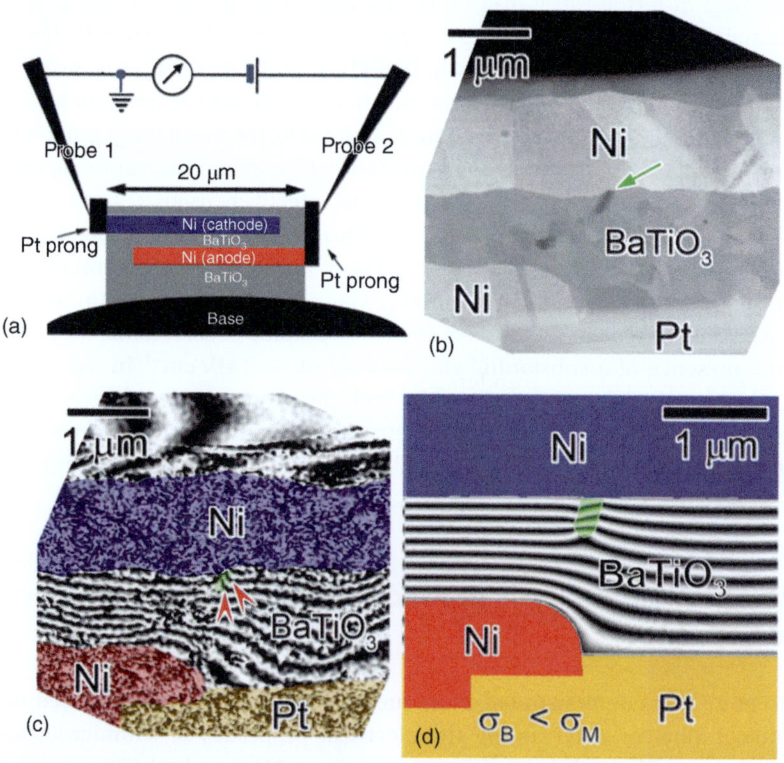

Figure 7.5 (a) Schematic of a thin-foil specimen of a multilayer ceramic capacitor. (b) ADF-STEM and (c) reconstructed phase images around a precipitate indicated by an arrow and arrowheads. The phase information is amplified by a factor of two in the reconstructed phase image in (c). (d) Electric field distribution images obtained by simulation for the relationship $\sigma_B < \sigma_M$ between the electrical conductivities of the precipitate and the $BaTiO_3$ matrix.

shows that this precipitate is rich in Cr and poor in Si. Figure 7.5c is a reconstructed phase image taken under an effective applied voltage of 7 V; thus, the authors subtracted the phase-shift information acquired under an applied voltage of 3 V from that acquired under 10 V to suppress unwanted phase information such as substantial changes in the mean inner potential at the positions of the precipitates, thickness variations in the specimen, and the magnetic field from the Ni electrodes. Notably, changes occurred in both the spacing and directions of the contour lines in the local area of the precipitate, which the authors attributed to the difference in the electrical conductivity between the precipitate and the $BaTiO_3$ matrix. Figure 7.5d shows a two-dimensional simulated electric field map based on finite element method (FEM) modeling. In this model, the electrical conductivity of the precipitate σ_B is substantially smaller than the conductivity of the matrix σ_M. The results in Figure 7.5d explain several features of the observations in Figure 7.5c, such as the reduced spacing of the contour lines within the precipitate and the gradual change in the spacing in the matrix area. The conductivity of the Cr-rich phase should be smaller than that of the matrix by two or three orders of magnitude. This method can lead to a deeper understanding of the relationship between the complex microstructure and the material functionalities in capacitors widely used in industry.

7.3 Analysis of Spontaneous Polarization in Oxide Heterojunctions

Electron holography can be applied to characterize oxide heterojunctions in which spontaneous polarization plays a dominant role in the charge screening. The materials studied are trivalent transition-metal-oxide insulators, which do not form a conductive interface with STO ($SrTiO_3$) even though the interface has a polar discontinuity [36]. Among them, $LaFeO_3$ (LFO), which is known as a centrosymmetric Mott insulator with an optical gap (2.2 eV) in the visible-light region, is one of the most suitable materials [37]. It exhibits a high resistivity compared with other Mott insulators because of the closed-shell structure in the up-spin band of the Fe^{3+} state [38]. In addition, it exhibits instability toward the polar state under epitaxial strain [39]. Thus, the LFO=STO junction is as an attractive model system to examine interface-induced spontaneous polarization. We deposited 30-nm-thick LFO films on 0.5-mm-thick STO and $SrTi_{0.9998}Nb_{0.0002}O_3$ (Nb:STO) substrates with controlled TiO_2 and SrO terminations, as illustrated in the left of Figure 7.6a [40].

Figure 7.6a shows the atomic structure and elemental distributions near the LFO=STO interfaces imaged by STEM and EDS. The image verifies that the elemental sequence was constructed as we designed. The broadening of the distribution estimated from the STEM–EDS line profiles of the intensities remains, at most, 2 unit cells from the interface [40], indicating that the interface is sufficiently abrupt for the interfacial atomic sequence to be classified into two definite types and that the effect of interfacial disorder is negligible. The absence of lateral conductivity in the LFO=STO junctions irrespective of the interface terminations is confirmed.

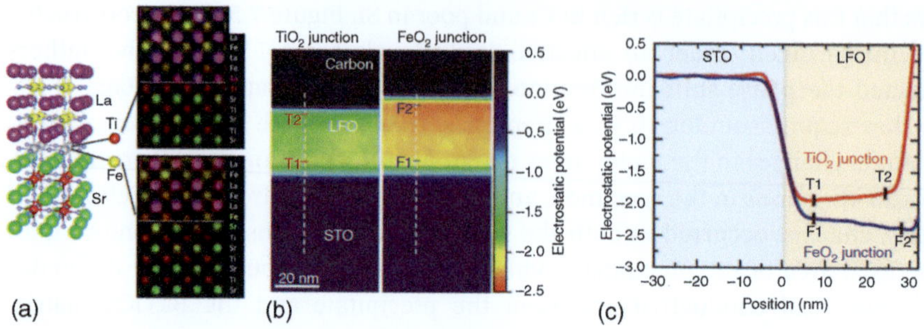

Figure 7.6 (a) Elemental distribution mappings for a TiO_2 junction (upper panel) and FeO_2 junction (lower panel) taken by STEM, where atomic-resolution EDS images are superimposed on HAADF images. (b) Mappings of the electrostatic potentials derived from electron holography for two LFO/STO junctions with different interface terminations. (c) Line profiles of the potentials along the dashed lines in (b).

Electron holography can directly quantify the spatial distribution of the electrostatic potential and, accordingly, the built-in internal electric field in a specimen [40]. Using this technique, we examined the possible origin of the electric polarization. Figure 7.6b shows the mappings of the electrostatic potentials for an electron, where we denote the junctions formed on the TiO_2- and SrO-terminated STO surfaces as the "TiO_2 junction" and the "FeO_2 junction" according to the interfacial transition-metal layer, respectively. The electrostatic potential is almost uniform in the in-plane direction in both junctions. We can now examine in detail the electrostatic potential; Figure 7.6c shows two examples of the electrostatic potential obtained along the dashed lines in Figure 7.6b.

Thus, electron holography can reveal that the spontaneous polarization dominates the electric polarization. The polarization direction can be switched by changing the chosen interfacial atomic sequence. It is noted that the emergent spontaneous polarization induces anomalous photovoltaic properties in the junctions, such as a sign reversal in the photocurrent depending on the interfacial atomic sequence and the excitation photon energy.

7.4 Evaluation of Electric Charge with Laser Irradiation

We here briefly review studies in which the amount of charge in nonconductive materials was successfully determined. Frost et al. [41] acquired electron holograms from charged latex spheres and compared the reconstructed phase images, which represented the equipotential lines outside the charged sphere, with results obtained by calculations based on the classical method of electromagnetism. On the basis of this comparison, they determined the number of electrons responsible for the charging. Hirayama et al. [42] performed a unique experiment using two biprisms attached to a transmission electron microscope to examine the charging effect in alumina and latex particles. On the basis of the obtained three-wave interference patterns, they quantitatively determined the amount of charge.

Here, we describe analyses of the amount of charge in commercial organic photoconductors, which are widely used in electrophotography. We conducted in situ observations of triboelectricity and subsequent discharge events using a double-probe piezodriving holder to which a laser port was attached, as explained in Section 6.6. Although the experimental result is preliminary, we expect that this method can be used to evaluate the efficiency of charge generation depending on the materials and the power of the laser beam. Note that the discharge effect in organic photoconductors has not yet been directly analyzed. In most laser printing systems, electrostatic latent charges are produced through the so-called corona discharge effect on the surface of the organic photoconductor, which comprises a charge-transport layer, a charge-generation layer, an undercoat layer, and a substrate. Instead of using the corona discharge effect to generate the electrostatic latent charges, we used frictional charging via a small glass rod (Figure 7.7a).

To avoid charging caused by the emission of electron-induced secondary electrons, we placed an electric shield above the specimen in a manner similar to that in the previous experiment on toner particles, as described in Section 6.6. We directly fixed a Mo plate at the top of the organic photoconductor, as represented by the "shield" in Figure 7.7a. After frictional charging using the glass rod, negative charges were expected to appear on the surface of the organic photoconductor (Figure 7.7b). To investigate the formation of negative charges, electron holograms were observed. The specimen was subsequently irradiated with a laser (Figure 7.7c), to reduce the density of the negative charges.

Figure 7.7d shows the reconstructed phase image obtained before the laser irradiation. The figure explicitly indicates the presence of negative charges beneath the shield, implying the formation of negative triboelectric charges in the specimen,

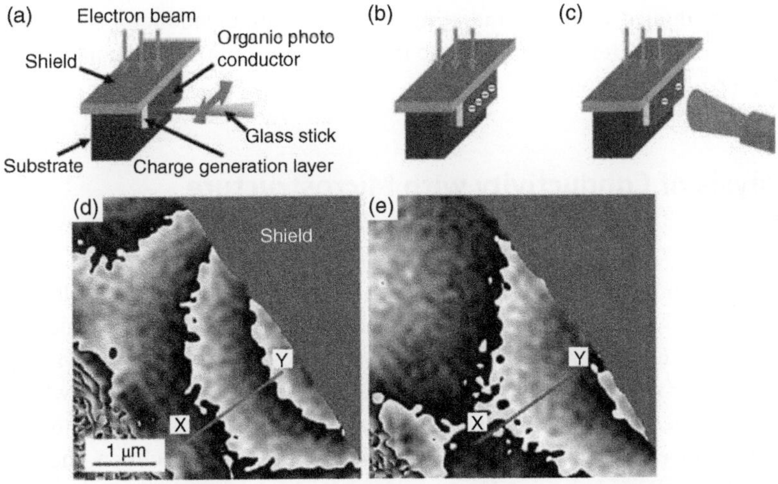

Figure 7.7 (a–c) Schematic of the process of charging through mechanical friction and the subsequent discharge by laser irradiation. Reconstructed phase images obtained (d) before and (e) after the laser irradiation.

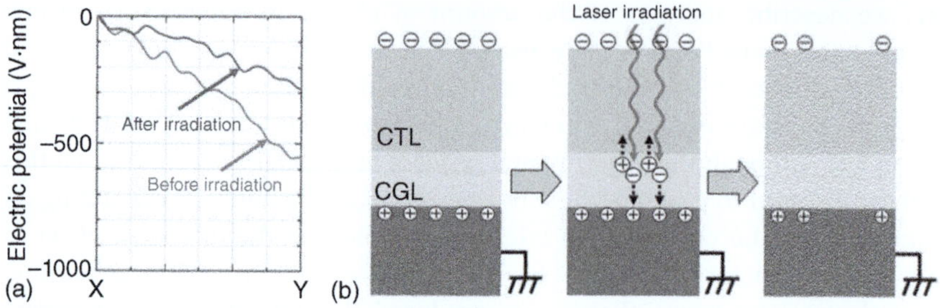

Figure 7.8 (a) Laser irradiation-induced change in the electric potential distribution, projected onto the incident electron beam. Electric potential at point X is set to 0. (b) Schematic of the mechanism of negative charge reduction on an organic photoconductor upon laser irradiation. CTL, charge-transport layer; CGL, charge-generation layer.

which are not visible in the figure. The spacing of the contour lines is markedly reduced by the laser irradiation, as demonstrated in Figure 7.7e.

In Figure 7.8a, the electric field caused by the negative charges around the organic photoconductor is reduced because of the laser irradiation (Figure 7.7c). Some of the negative charges might have disappeared because of electron–hole annihilation, which involves the generation of holes in the charge-generation layer of the organic photoconductor as a result of the laser irradiation (Figure 7.8b). The charge amount could be quantitatively analyzed by comparing the observed phase images with simulated images [43]. Thus, we demonstrated that the double-probe piezodriving holder with a laser irradiation port is suitable for investigating photoinduced phenomena, such as the discharge effect of laser irradiation on organic photoconductors.

In addition, through computer simulation, we quantitatively analyzed the electric field. As clarified again in the present study, the change in the electric charge amount can be evaluated without irradiating an organic photoconductor specimen with incident electrons. Here, we again note the importance of the field concept, as cited in Chapter 1 and Section 6.6.

7.5 Analysis of Conductivity with Microstructure Changes

By using a probe in the piezodriving specimen holder and applying an electric current, we can analyze changes in both the electric field and the microstructure to understand the characteristic conductive properties. Figure 7.9 shows the current–voltage (I–V) characteristics of a Ag-based conductive adhesive [44], which has attracted considerable attention as an alternative to Pb-free solder [45, 46]. In the figure, when the electric current is first supplied to the cured specimen, the I–V curve deviates from ohmic behavior. However, in the subsequent measurement, the I–V curve is approximately linear. Figure 7.10a illustrates a circuit used to supply a

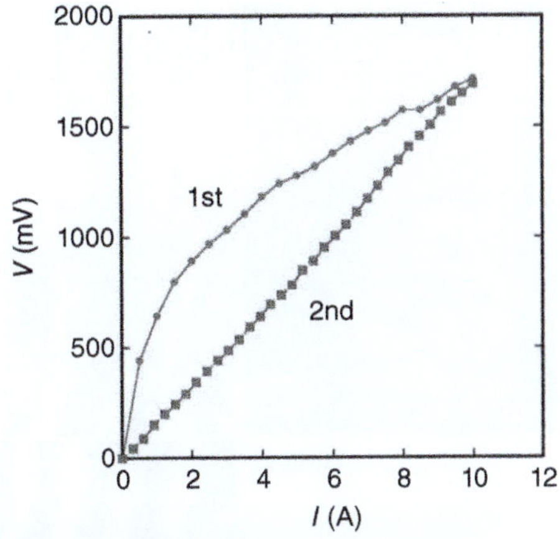

Figure 7.9 Nonlinear I–V curve measured for an initial state of cured bulk Ag-based conductive adhesive. Typically, I–V curve changes to exhibit ohmic behavior in a second measurement.

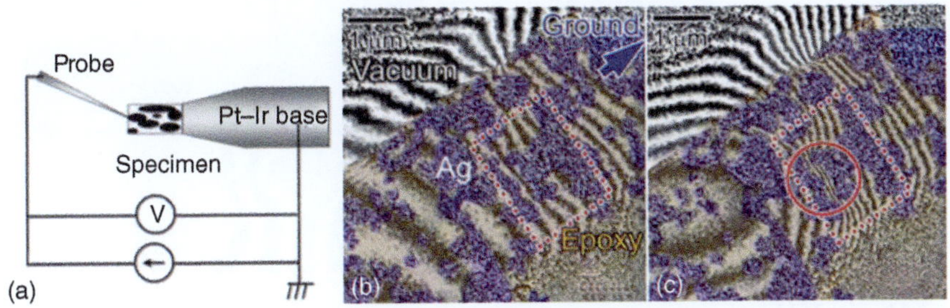

Figure 7.10 (a) Experimental setup for electric field analysis of a Ag-based conductive adhesive. Reconstructed phase images obtained (b) before and (c) after a large electric current of 1 μA is supplied. Source: Kawamoto et al. [44], with permission from AIP Publishing LLC.

constant current to a thin-foil Ag-based conductive adhesive attached to the needle tip of the Pt–Ir base.

Figure 7.10b shows the reconstructed phase image obtained at an applied voltage of 3 V. The violet and orange portions indicate the Ag filler and the epoxy, respectively. The effect of the mean inner potential was eliminated by subtracting the phase information obtained under a zero-voltage condition from that obtained at an applied voltage of 3 V. Because of the large thickness of the specimen (300 nm), observations inside the Ag agglomerations were not possible. Nevertheless, in the epoxy area, equipotential lines approximately parallel to each other are clearly observed. Figure 7.10c shows a reconstructed phase image of the same field of view after a large electric current of 1 μA was supplied. Interestingly, in Figure 7.10c, the spacing of the contour lines is markedly decreased, as indicated by the circle.

The change in the electric field can be understood on the basis of the change in conductivity in a local area. Figure 7.11a shows a magnified TEM image of the area indicated by the dotted squares in Figure 7.10b,c. The four Ag agglomerations are labeled as A, X, Y, and Z. Figure 7.11b,c shows reconstructed phase images of

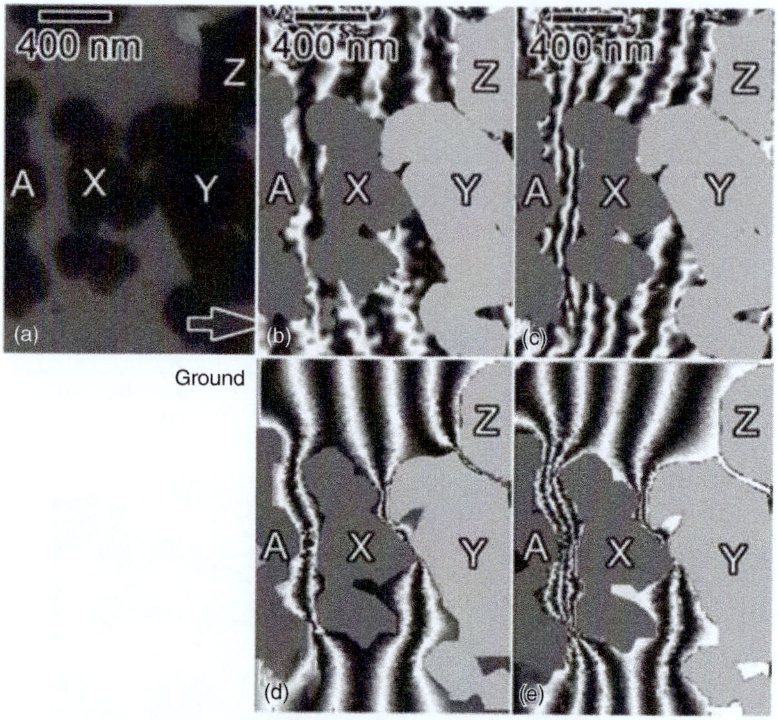

Figure 7.11 (a) TEM image corresponding to area indicated by dotted lines in Figure 7.10b,c. Reconstructed phase images obtained (b) before and (c) after a 1 μA current is supplied. (d and e) Simulations of the equipotential lines. Source: Kawamoto et al. [44], with permission from AIP Publishing LLC.

this area, as obtained before and after the large electric current of 1 μA was supplied, respectively. To simulate the potential distribution in this field of view, the agglomerations were modeled using elemental plates (300 nm in thickness) with similar shapes (Figure 7.11d,e). As evident in Figure 7.11d, the features observed in Figure 7.11b were satisfactorily reproduced by assuming the following electric potentials for the Ag agglomerations: $V_A = 1.2$ V, $V_X = 0.8$ V, $V_Y = 0.4$ V, and $V_Z = 0$ V. The features in Figure 7.11c were approximated using the following parameters: $V_A = 1.2$ V, $V_X = 0.45$ V, and $V_Y = V_Z = 0$ V. The condition $V_Y = V_Z$ may be consistent with our interpretation that the large electric current (1 μA) enhanced the conductivity in the local area between Ag agglomerations Y and Z by causing an irreversible morphological change [44].

The bright-field images in Figure 7.12a,b show the microstructure change before and after the large current was supplied, respectively. Slight changes apparently occurred in the morphology of the Ag agglomerations. For example, as indicated by the red and blue arrows in Figure 7.12b, tiny horns can be observed in the Ag agglomeration, whereas the surface of the same agglomeration is smooth in Figure 7.12a. Thus, this morphological change may affect the local conductivity of the specimen by changing the local electric field distribution. The phenomena

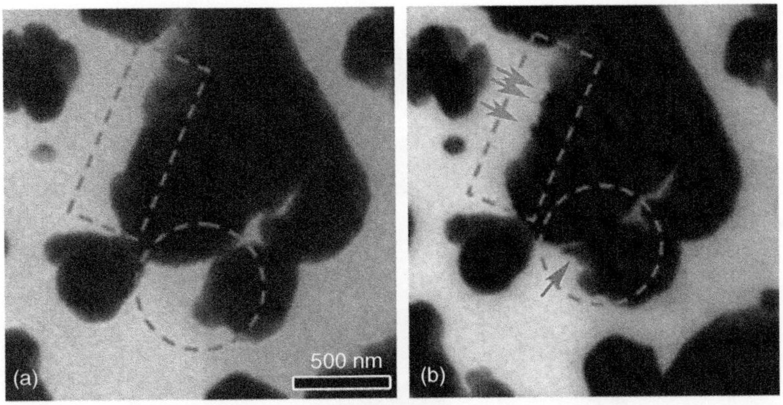

Figure 7.12 Bright-field images of the specimen (a) before and (b) after a 1 μA current is supplied. Source: Kawamoto et al. [44], with permission from AIP Publishing LLC.

observed with the model specimens of Ag-based conductive adhesive and the surface-morphology-dependent electric potentials were simulated to understand the local conductive mechanism [47]. Several researchers have reported that an electric current was passed even in the space between detached Ag fillers [48, 49], although the underlying mechanism has not been fully explained. In such a peculiar condition of the electric current, the shapes of the Ag fillers might be an important factor. Here, we note that the change in the morphology of Ag horns resulted from the electron current. We cannot, however, observe the change in electric field due to the electron current because this is too small for its field change to be detected. This point is further discussed in Part IV.

A similar morphological change was observed in model specimens. As shown in Figure 7.13b, a cathode Ag probe was placed near an anode Ag probe, with a spacing of ∼40 nm. Notably, this space was only partly filled with epoxy resin, as evident from the area of gray contrast representing the presence of epoxy resin in Figure 7.13b. Using this experimental setup, we repeated the *I*–*V* curve measurements nine times. Some of the results are plotted in Figure 7.13a. The image in Figure 7.13c shows the shape of an anode needle after the third *I*–*V* measurement. The anode needle was deformed (it appeared to have been attracted to the cathode needle), presumably because of coulombic interaction under the applied voltage. More importantly, the anode probe in Figure 7.13c showed a small horn, which was likely to form in the portion indicated by an arrow. This horn was still recognizable in Figure 7.13d after the ninth *I*–*V* measurement. Compared with the result in Figure 7.13c, the results in Figure 7.13d do not clearly show a horn. We attribute this difference to the effect of tilting in the anode needle with respect to incident electrons; i.e. the bright-field images in Figure 7.13b–d are projections of the model specimen. Interestingly, the shift of the *I*–*V* curves Figure 7.13a occurred concurrently with this irreversible change in the needle shape. These observations strongly indicate that the irreversibility of the *I*–*V* characteristics (e.g. shifting of the *I*–*V* curves) was closely related to the morphological change of the Ag electrodes. This result, which

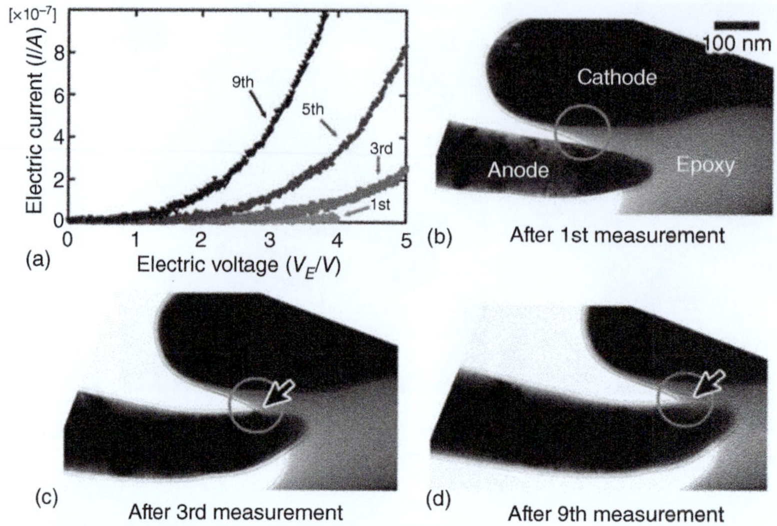

Figure 7.13 (a) Current–voltage (*I*–*V*) curves observed in a model specimen with an epoxy thickness of 40 nm. (b–d) TEM images of the model specimen, as observed after repeating the *I*–*V* measurement: (c) after the third measurement and (d) after the ninth measurement. Source: Kurosu et al. [47], with permission from IEEE.

was obtained using model specimens, is consistent with the results of Kawamoto et al. who directly observed thin-foil specimens of a commercial conductive adhesive [44].

As previously mentioned, the anode needle appears to be deformed slightly during repeated *I*–*V* measurements (i.e. the spacing between the anode and cathode appears to be reduced, as shown in Figure 7.13). This morphological change is consistent with the reduction of the critical voltage required for a steep increase in the *I*–*V* curves. In addition to focusing on this shape deformation, we also focus on the effect of the horn formed in a Ag electrode subjected to a large current on the local electric field distribution.

To calculate equipotential lines around the Ag electrodes, which resemble the cathode and anode needles in Figure 7.13b, we used the structural model shown in Figure 7.14a. Note that Figure 7.14a shows a model representing the original state (i.e. without horns). The other state involving a small horn is presented in Figure 7.14c, where the horn was located in the lower electrode (inside the rectangular red dashed line), as observed in Figure 7.13c. The height of the horn was fixed at 4 nm on the basis of the observation in Figure 7.13, although it was assumed to be round for simplicity. The area of epoxy resin, indicated by blue, was determined on the basis of TEM observations. The other portion was vacuum. The dimensions in the *x*–*y* plane were comparable with the dimensions of projected image in Figure 7.13b, although the thickness along the *z*-axis was assumed to be constant (100 nm) for both the Ag electrodes and the epoxy resin. The voltages of the upper

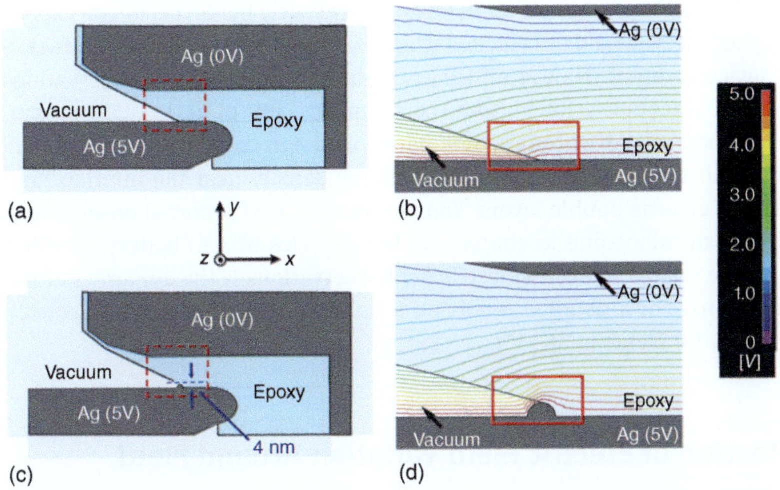

Figure 7.14 Simulations of local electric field of model specimen. (a) Structural model that includes no horns in the Ag electrode. (b) Equipotential lines calculated for area indicated by red-dotted lines in (a). (c) Structural model including a horn in lower Ag electrode. (d) Equipotential lines calculated for area indicated by red dotted lines in (c).

and lower electrodes were fixed at 0 and 5 V, respectively. Figure 7.14b shows the calculation result for the area enclosed by the dashed line in Figure 7.14a. The simulation indicates equipotential lines that reflect the shape of the electrodes. The line spacing was different between the epoxy resin and vacuum because of the difference in electric permittivity. Figure 7.14d shows the modulation of equipotential lines in the presence of a horn in the lower electrode. This modulation was substantial near the horn, around which the lines were highly bent (e.g. compare the areas indicated by the rectangles (red solid lines) in Figure 7.14b,c). More importantly, the line spacing near the horn was reduced compared with the result shown in Figure 7.14b. Our calculations estimated the density of electric charge (induced by the applied voltage) to be $7.4 \times 10^{-3}\ \mathrm{C\,m^{-2}}$ at the position of the horn in Figure 7.14d. This value is more than 20% larger than the result calculated for the same position on the lower electrode in Figure 7.14b. Because the spacing of the upper and lower electrodes was only 40 nm, the applied voltage (5 V) produced a substantial electric field on the order of $10^5\ \mathrm{V\,cm^{-1}}$. The results shown in Figure 7.13a indicate that the I–V curve was already in a state of steep increase under this strong electric field. Therefore, even a 20% modulation would contribute substantially to the enhancement of local conduction. As mentioned in the discussion of the previous TEM observations, the spacing between Ag agglomerations reached the order of 10 nm (i.e. comparable with the spacing in our simulation) in the commercial conductive adhesive. On the basis of these results, we propose that the formation of small horns, which was observed in cured Ag-based adhesives subjected to large currents, played an important role in causing the irreversible I–V characteristic.

Regarding electron holography studies on applied electric fields (i.e. in situ experiments on applied electric fields), Twitchett-Harrison et al. [50] analyzed biased semiconductor device samples. Han et al. [51] performed similar in situ observations with an improved geometry for biasing the samples. Tanji et al. [52, 53] investigated the inner potential distribution at a metal–oxide-ion conductor interface and demonstrated substantial modulation of the charge density near the interface and the formation of electric double layers. Yamamoto et al. [54] reported observations of electric field variations due to charge–discharge cycles in a Li battery. Another challenging topic related to in situ electric field observations is the dynamics of ferroelectric domains. For a detailed report on a holographic study of ferroelectrics, refer to the paper by Lichte et al. [55].

7.6 Detection of Electric Field Variation Around Field Emitter

To understand the emission properties of field emitters, evaluating the electric potential around them is important. In particular, for electron holography studies, an electron beam with high coherence can be produced from a field-emission gun (FEG). Regarding the observation of emitters or metallic tips, Matteucci et al. [56] performed an incipient study of the potential distribution near a tungsten microtip. Cumings et al. [57] showed a reconstructed phase image representing the field emission from a carbon nanotube. We herein report our observations on the topic of emitters.

Figure 7.15a–d shows electron holograms (upper panels) and reconstructed images (lower panels) obtained from unused (a and c) and used (b and d) Schottky emitters at an applied voltage of 100 V [58]. We observe a difference in the edge

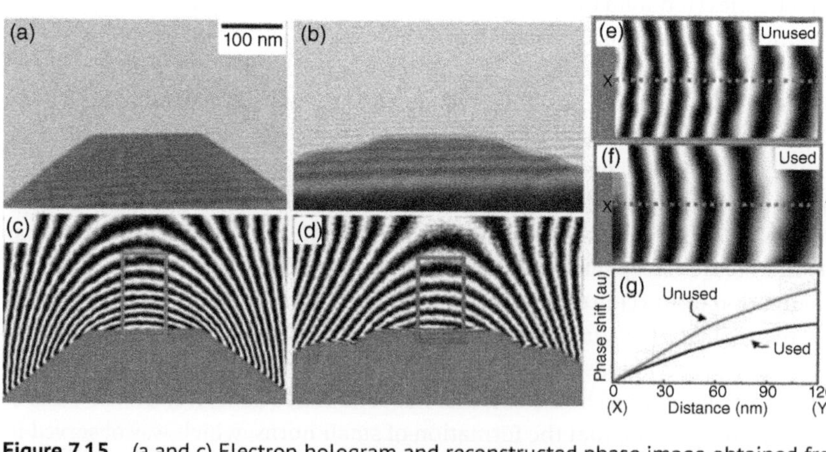

Figure 7.15 (a and c) Electron hologram and reconstructed phase image obtained from an unused emitter; (b and d) electron hologram and reconstructed phase image obtained from a used emitter; (e and f) enlarged images of rectangular areas shown in (c) and (d), respectively; and (g) plots of phase shifts measured along each line $X-Y$. Source: Tetsuo Oikawa et al. (2007), with permission from Oxford University Press.

shapes – specifically, a deformation in the W(100) facet of the used emitter. This observation implies that sputtering effects of the residual ions under the influence of the large electric field caused the W(100) facet to wear. Accordingly, the electric potential differs around the field emitters, as shown in Figure 7.15e,f, which are enlarged images of the areas marked in Figure 7.15c,d, respectively. The difference in the electric potential distribution is clear in Figure 7.15g, which shows plots of the phase shift along each line X–Y. Variations in the electric potential distribution around a cold-type FEG emitter were also studied using a similar analysis method [59].

Figure 7.16a shows the experimental setup and the obtained I–V curve used to study the field emission from a single $TaSi_2$ nanowire [60]. The I–V curve was recorded using a transmission electron microscope. When the voltage increased to approximately 50 V, a certain emission current (~1 nA) was detected. The reconstructed phase images in Figure 7.16b show the change in the electric potential near the $TaSi_2$ nanowire with increasing applied voltage. The observations represent the

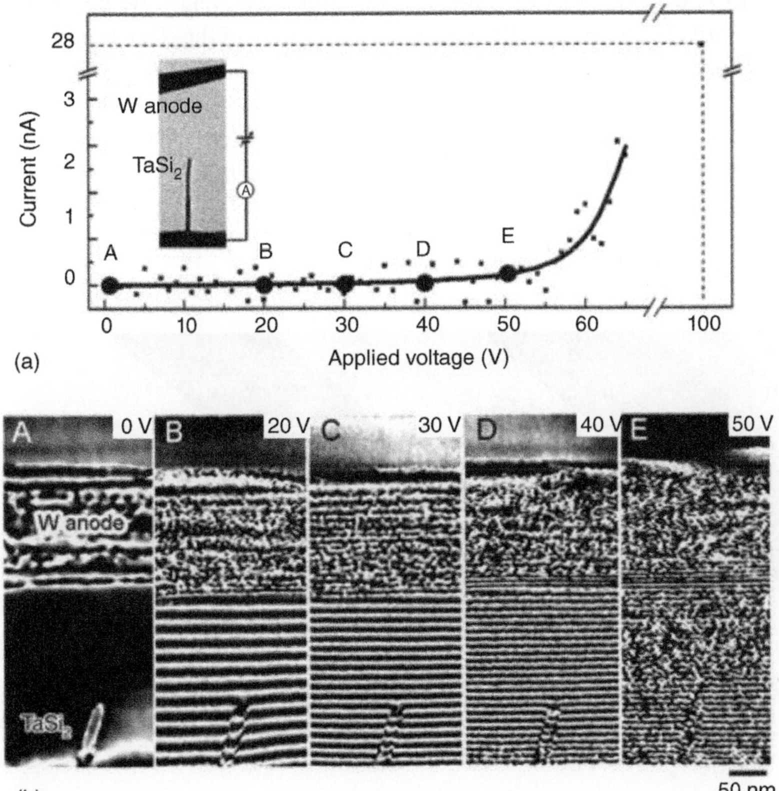

Figure 7.16 (a) I–V curve showing field emission from a $TaSi_2$ nanowire, observed using the experimental setup shown in the inset and (b) the changes in the reconstructed phase images with increasing applied voltage. Source: Kim et al. [60], with permission from American Chemical Society.

electric field projected along the incident electron beam. With increasing applied voltage, the spacing between the equipotential lines decreases. The reconstructed phase image is highly blurred at a voltage of 50 V. Note that the equipotential lines shown in Figure 7.16b are approximately parallel to the anode edge: the peculiar modulation of contour lines is difficult to observe because of the charged emitters, as previously demonstrated by Matteucci et al. [56] and Cumings et al. [57]. This phenomenon is observed because, in this experiment, a large electric field was applied between the anode and cathode, which were both substantially larger than the nanowire. The phase shift due to the tiny nanowire, which protrudes from the large cathode, should be much smaller than that due to the anode and cathode. That is, the contribution of the large cathode is so strong that the phase shift due to the nanowire is unclear in Figure 7.15.

Figure 7.17a,b shows the reconstructed phase image obtained at an applied voltage of 50 V and a simulated image, respectively. In Figure 7.17a, a region of irregular contrast is observed between the W anode and the $TaSi_2$ nanowire, as outlined in green. The simulated image was obtained by assuming a bias voltage of 50 V, with a fluctuation of 1 V. By comparing the observed and simulated images, we concluded that the irregularity in contrast resulted from the fluctuation of the electric potential in the nanowire, as a result of ballistic emission [60].

Although the detailed electric field around the field emitter was clarified by applying a bias voltage, the electric field variation due to the motions of electrons emitted from the field emitters could not be detected because the electric field variation due to the emission current of ~0.2 nA was too small to detect. This point is discussed further in Part 4, where we describe the collective motions of secondary electrons observed around charged insulators.

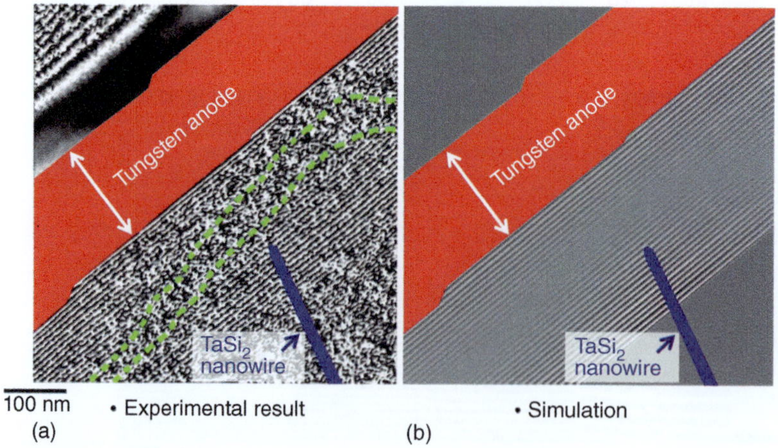

Figure 7.17 (a) Reconstructed phase image obtained from a $TaSi_2$ nanowire at an applied voltage of 50 V and (b) a simulation result for comparison.

References

1 Joy, D.C., Roming, A.D., and Goldstein, J. (1986). *Principles of Analytical Electron Microscopy*. Springer Science & Business Media.
2 Reimer, L. (1984). *Transmission Electron Microscopy: Physics of Image Formation and Microanalysis*. Springer.
3 Egerton, R.F. (1996). *Electron Energy-Loss Spectroscopy in the Electron Microscope*, 2e. New York: Plenum.
4 Williams, D.B. and Carter, C.B. (1996). *Transmission Electron Microscopy: A Textbook for Materials Science*. Springer.
5 Kelly, P.M., Jostsons, A., Blake, R.G., and Napier, J.G. (1975). *Phys. Status Solidi A* 31: 771.
6 Allen, S.M. (1981). *Philos. Mag. A* 43: 325.
7 Tanaka, M. and Terauchi, M. (1985). *Convergent-Beam Electron Diffraction*. Tokyo: JEOL LTD.
8 Spence, J.C.H. and Zuo, J.M. (1992). *Electron Microdiffraction*. New York: Plenum Press.
9 Nishino, D., Nakafuji, A., Yang, J.-M., and Shindo, D. (1998). *ISIJ Int.* 38: 1369.
10 Shindo, D. and Akase, Z. (2020). *Mater. Sci. Eng. R Reports* 142: 100564. https://doi.org/10.1016/j.mser.2020.100564.
11 Shindo, D., Musashi, T., Ikematsu, Y. et al. (2005). *J. Electron Microsc.* 54: 11.
12 Rau, W.D., Schwander, P., Baumann, F.H. et al. (1999). *Phys. Rev. Lett.* 82: 2614.
13 Wang, Z., Hirayama, T., Sasaki, K. et al. (2002). *Appl. Phys. Lett.* 80: 246.
14 Yoo, J.H., Yang, J.-M., Ulugbek, S. et al. (2008). *J. Electron Microsc.* 57: 13.
15 Twitchett-Harrison, A.C., Yates, T.J.V., Newcomb, S.B. et al. (2020). *Nano Lett.* 2007: 7.
16 Wolf, D., Lubk, A., Lichte, H., and Friedrich, H. (2010). *Ultramicroscopy* 110: 390.
17 den Hertog, M.I., Schmid, H., Cooper, D. et al. (2009). *Nano Lett.* 9: 3837.
18 Ichikawa, S. (2010). *Microscopy* 45: 147.
19 Wolf, D., Lubk, A., Röder, F., and Lichte, H. (2013). *Curr. Opin. Solid State Mater. Sci.* 17: 126.
20 Ikarashi, N., Takeda, H., Yako, K., and Hane, M. (2012). *Appl. Phys. Lett.* 100: 143508.
21 Yazdi, S., Kasama, T., Beleggia, M. et al. (2015). *Ultramicroscopy* 152: 10.
22 Han, M.-G., Marshall, M.S.J., Wu, L. et al. (2014). *Nat. Commun.* 5: 4693.
23 Migunov, V., London, A., Farle, M., and Dunin-Borkowski, R.E. (2015). *J. Appl. Phys.* 117: 134301.
24 Akase, Z. and Shindo, D. (2019). *Mater. Trans.* 60: 2120.
25 Cai, J. and Ponce, F.A. (2002). *J. Appl. Phys.* 91: 9856.
26 Li, L., Ketharanathan, S., Drucker, J., and McCartney, M.R. (2009). *Appl. Phys. Lett.* 94: 232108.

27 Marino, F.A., Cullen, D.A., Smith, D.J. et al. (2010). *J. Appl. Phys.* 107: 054516.

28 Kawamoto, N., Ono, H., Terasaki, Y. et al. (2019). *Mater. Trans.* 60: 2109.

29 Kishi, H., Okino, Y., Honda, M. et al. (1997). *Jpn. J. Appl. Phys.* 36: 5954.

30 Polotai, A.V., Jeong, T.-H., Yang, G.-Y. et al. (2007). *J. Electroceramics* 18: 261.

31 Metzmacher, C. and Albertsen, K. (2001). *J. Am. Ceram. Soc.* 84: 821.

32 Murakami, Y., Kawamoto, N., Shindo, D. et al. (2006). *Appl. Phys. Lett.* 88: 223103.

33 Kawamoto, N., Murakami, Y., Shindo, D. et al. (2009). *Mater. Trans.* 50: 1572.

34 Kawamoto, N., Kakefuda, Y., Yamada, I. et al. (2018). *Nano Energy* 52: 323.

35 Kuramae, R., Ono, H., Fujikawa, Y. et al. (2012). *Mater. Trans.* 53: 696.

36 Chambers, S.A., Qiao, L., Droubay, T.C. et al. (2011). *Phys. Rev. Lett.* 107: 206802.

37 Imada, M., Fujimori, A., and Tokura, Y. (1998). *Rev. Mod. Phys.* 70: 1039.

38 Arima, T., Tokura, Y., and Torrance, J.B. (1993). *Phys. Rev. B* 48: 17006.

39 Zhao, H.J., Yang, Y., Ren, W. et al. (2014). *J. Phys. Condens. Matter.* 26: 472201.

40 Nakamura, M., Kagawa, F., Tanigaki, T. et al. (2016). *Phys. Rev. Lett.* 116: 156801.

41 Frost, B.G., Allard, L.F., Völkl, E. et al. (ed.) (1995). *Electron Hologr*, 169. Amsterdam: North-Holland.

42 Shindo, D. and Oikawa, T. (2002). *Analytical Electron Microscopy for Materials Science*, 111. Springer.

43 Shindo, D., Takahashi, K., Murakami, Y. et al. (2009). *J. Electron Microsc.* 58: 245.

44 Kawamoto, N., Murakami, Y., and Shindo, D. (2010). *J. Appl. Phys.* 107: 044309.

45 Mir, I. and Kumar, D. (2008). *Int. J. Adhes. Adhes.* 28: 362.

46 Yim, M.J., Li, Y., Moon, K. et al. (2008). *J. Adhes. Sci. Technol.* 22: 1593.

47 Kurosu, K., Kawamoto, N., Murakami, Y., and Shindo, D. (2012). *IEEE Trans. Comp. Packag. Manuf. Technol.* 2: 294.

48 Lu, D., Tong, Q.K., and Wong, C.P. (1999). *IEEE Trans. Electron. Packag. Manuf.* 22: 228.

49 Nakamura, S., Saito, K., Sawa, G., and Kitagawa, K. (1997). *Jpn. J. Appl. Phys.* 36: 5163.

50 Twitchett, A.C., Dunin-Borkowski, R.E., and Midgley, P.A. (2002). *Phys. Rev. Lett.* 88: 238302.

51 Han, M.-G., Smith, D.J., and McCartney, M.R. (2008). *Appl. Phys. Lett.* 92: 143502.

52 Tanji, T. (2010). *Microscopy* 45: 152.

53 Tanji, T., Moritomo, H., Shimura, T. et al. (2007). *Microsc. Microanal.* 13: 1224.

54 Yamamoto, K., Iriyama, Y., Asaka, T. et al. (2010). *Angew. Chem. Int. Ed.* 49: 4414.

55 Lichte, H., Reibold, M., Brand, K., and Lehmann, M. (2002). *Ultramicroscopy* 93: 199.

56 Matteucci, G., Missiroli, G.F., Muccini, M., and Pozzi, G. (1992). *Ultramicroscopy* 45: 77.

57 Cumings, J., Zettl, A., McCartney, M.R., and Spence, J.C.H. (2002). *Phys. Rev. Lett.* 88: 056804.

58 Oikawa, T., Kim, J.J., Tomita, T. et al. (2007). *J. Electron Microsc.* 56: 171.

59 Kim, J.J., Xia, W.X., Shindo, D. et al. (2007). *Mater. Trans.* advpub: 0709030182.

60 Kim, J.J., Shindo, D., Murakami, Y. et al. (2007). *Nano Lett.* 7: 2243.

8

Magnetic Field Analysis

8.1 Quantitative Analysis of Magnetic Flux Distribution of Nanoparticles

Tripp et al. [1] demonstrated a stable flux-closure state in bracelet-like rings of Co particles (average diameter: 27 nm) at room temperature, meaning that the rings were created by dipole-directed self-assembly. Sugawara et al. [2] investigated the temperature dependence of the magnetic order parameter in bracelet-like rings of Ni particles (diameter: 30 nm) and reported a rapid decrease in the parameter upon heating, as compared with the result for a bulk Ni specimen. Hÿtch et al. [3] examined the magnetic vortex structure achieved in chained Fe–Ni particles (average diameter: 50 nm). They reported that the diameters of the vortex cores depended on their orientation with respect to the chain axis and that the vortex formation could be controlled by the presence of smaller particles in the chains. Gao et al. [4] reported a unique spin-order structure in self-assembled Co nanodisks (diameter: 18 nm; thickness: 5 nm), wherein a spiral spin arrangement along the row axis was achieved.

Gao et al. [5] also observed chains composed of core–shell Co–CoO nanocrystals (Figure 8.1) and reported that the magnetic thermal stability of the Co nanocrystals was remarkably improved by the exchange coupling between the ferromagnetic (FM) core and the antiferromagnetic (AFM) shell. Although Figure 8.1 reveals the magnetic microstructure of the chained magnetic particles, in which the lines of magnetic flux exist mainly inside the specimen (except at the end of the chain), observing an isolated magnetic particle requires special attention. In an isolated particle having the form of a single domain, a stray magnetic field with a direction opposite the magnetization vector exists above and below the particle. Biskupek et al. [6] demonstrated that this stray magnetic field reduced the observable phase shift in an isolated spherical particle by a factor of $2/\pi$.

Figure 8.2 shows analysis results for the magnetic structure of Co–CoO nanoparticles. To separate information related to the electric field (mean inner potential) and magnetic field, reconstructed phase images corresponding to two magnetic remanent states were observed by applying the magnetic field in both the left-to-right and right-to-left directions. From the difference between the two reconstructed phase images (Figure 8.2b), we obtained the electric field indicated

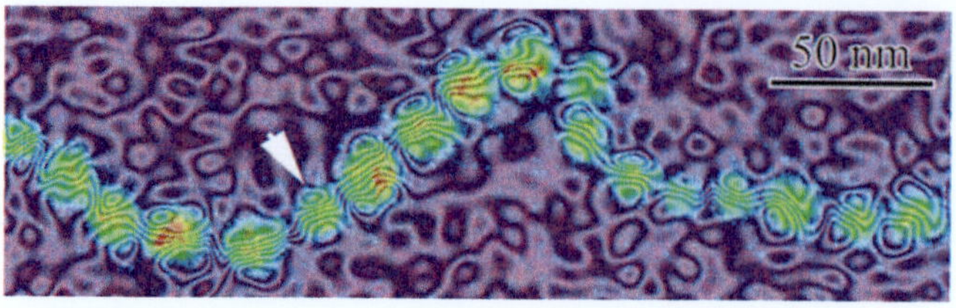

Figure 8.1 Magnetic microstructure of a chain composed of core–shell Co–CoO nanocrystals. Source: Gao et al. [5], with permission from AIP Publishing LLC.

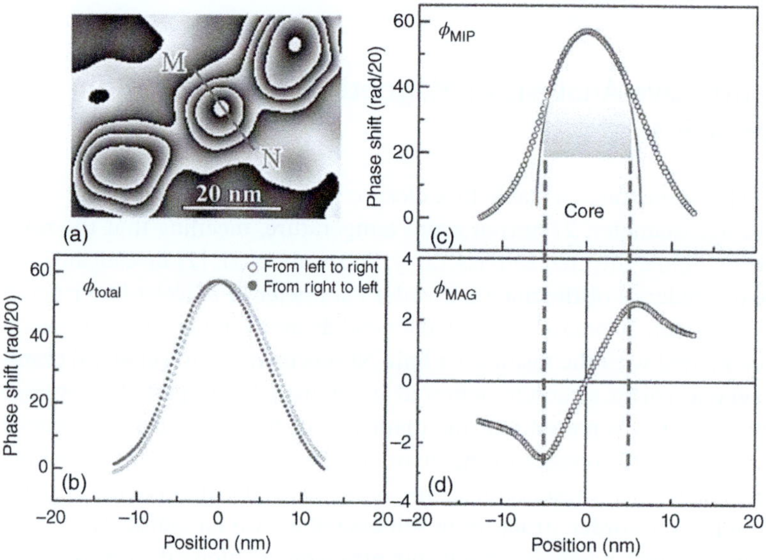

Figure 8.2 (a) Reconstructed phase image of the nanocrystal indicated in Figure 8.1. For line M–N across the nanocrystal, plots were constructed for (b) total phase shift ϕ_{total}, (c) contribution of the mean inner potential ϕ_{MIP}, and (d) contribution of the magnetic induction ϕ_{MAG}. Here, ϕ_{total} was observed in two remanence states, which were saturated along the left-to-right and right-to-left directions. Diameter of Co core and thickness of CoO shell were 11.2 and 4.9 nm, respectively. Mean inner potential of core was estimated to be 18.7 V, whereas that of shell was ~16.1 V. Solid curve in ϕ_{MIP} is theoretical fit curve.

by ϕ_{MIP} in Figure 8.2c. Along line M–N in (a), a magnetic field corresponding to a remanent state was obtained (Figure 8.2d). Continuous contours are observed running through the whole particle chain, suggesting that the nanocrystal spins are aligned head to tail along the chain direction. This state is one possible ground state for the arrangement of magnetic nanocrystals dominated by magnetic dipolar interactions. The curved chain is likely the result of magnetic dipolar interaction influenced by a collective Brownian motion in solution and/or additional shear forces during the drying process.

Figure 8.3 shows analysis results for the magnetic structure of Fe_3O_4 nanoparticles [7]. The magnetic field information shown in Figure 8.3a was extracted from two reconstructed phase images. One image was obtained after the specimen was flipped

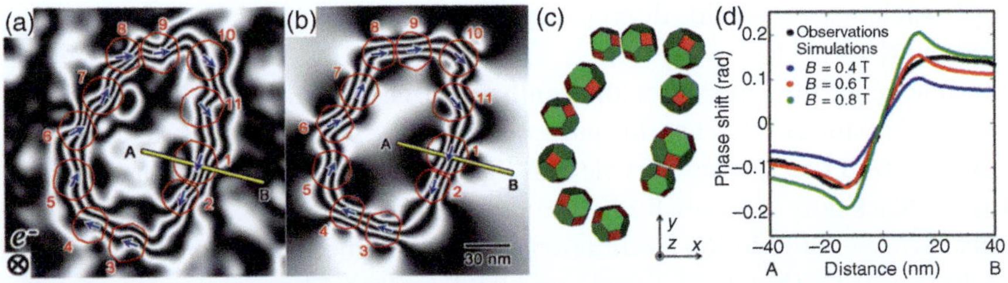

Figure 8.3 (a) Reconstructed phase image showing magnetic information of a ring-shaped self-assembly of Fe_3O_4 nanoparticles. Phase information was amplified by a factor of 100. Blue arrows and red lines indicate traces of <111> axis and outlines of nanoparticles, respectively. (b) Simulated reconstructed phase image showing magnetic signals. (c) Structure model used in simulation. (d) Comparison between observed (black) and simulated (blue, red, green) phase shifts observed along line A–B in (a) and (b). Simulation assumed three values for magnetic flux density (B) of Fe_3O_4 nanoparticles: 0.4 T (blue), 0.6 T (red), and 0.8 T (green). Phase information was not amplified. Source: Takeno et al. [7], with permission from AIP Publishing LLC.

so that it was upside down in reference to the incident electrons (see Section 4.3.4 [8]). For this observation, the sense of the phase shift was reversed for the magnetic field, although it remained unchanged for the electric field. A subtraction protocol using the two reconstructed phase images erased the undesired contribution of the electric field and accordingly revealed the magnetic field (Figure 8.3a). This figure was subjected to noise filtering, and the phase information was amplified by a factor of 100. The red lines and blue arrows indicate the outlines of individual nanoparticles and a plausible trace of the easy magnetization axis, respectively. The observation clearly shows the lines of magnetic flux, i.e. the contour lines tracing the ring.

Figure 8.3b shows a simulation result for the reconstructed phase image. As shown in Figure 8.3c, the nanoparticles were approximated by truncated octahedra, showing major facets of the {111} planes (indicated in green) and minor facets of the {100} planes (indicated in red). The size, location, and crystal orientation of the nanoparticles were fitted to the TEM observations [7]. Under the assumption that the magnetic flux density of the nanoparticles was 0.6 T [9, 10], we calculated the three-dimensional magnetic field, both inside and outside the particles, using the commercial program ELF/MAGIC [11]. The simulation reasonably explains the characteristic features of the observations. For paired particles 1–2 and 3–4, the traces of the easy magnetization axis (blue arrows) are almost parallel. For these particle pairs, the magnetic flux lines are approximately parallel to the blue arrows in both the observed (Figure 8.3a) and simulated (Figure 8.3b) results. By contrast, the flux lines are deformed in a zigzag manner for particle arrays 5–6–7–8 and 10–11–1. In fact, at these locations, the blue arrows are neither mutually parallel nor do they trace the ring. From these observations, we can attribute the fluctuations in the contour lines to the magnetocrystalline anisotropy, which is preserved in the ring-shaped self-assembly. Note that a small discrepancy is observed in the shape of the flux lines between the observed and simulated results, as in the case of particle

9. A plausible reason for this discrepancy is an additional phase shift due to Bragg reflection [7].

The magnetic flux density for the synthesized Fe_3O_4 nanoparticles can be also determined quantitatively. For this purpose, the phase shift was measured along the yellow line A–B that crosses particle 1 in Figure 8.3a. These observations are plotted as the black dots in Figure 8.3d. The other curves in Figure 8.3d show simulated results corresponding to the assumption that the magnetic flux density (B) in the particles was 0.4 T (blue), 0.6 T (red), and 0.8 T (green). These plots suggest that a flux density of 0.6 T best fits the observations. Although a small deviation (0.03 rad, between the black and red curves) is also observed, this value is comparable with the uncertainty in the phase analysis (0.03 rad), which was deduced to result from background noise. Assuming that the magnetic flux density of the nanoparticles is 0.6 T, this value corresponds to the flux density observed for bulk Fe_3O_4 crystals [9, 10]. Thus, within the accuracy of this electron holography study, we could not obtain any signatures indicating spin disorder in the neighborhood of the surface, which would suppress the effective magnetic flux density in the nanoparticles.

8.2 Observation of Magnetization Processes

8.2.1 Soft Magnetic Materials

For a typical soft magnetic material, such as a permalloy, the magnetic flux distribution can be observed in detail by electron holography. On the basis of detailed magnetic flux distributions in reconstructed phase images, Néel walls were reported to be visible in a relatively thin specimen, whereas cross-tie walls were observed in a thicker specimen [12]. All the soft magnetic materials presented in this section were thinned from bulk materials. Thus, magnetic structures in these thin specimens differ from those of the corresponding bulk materials. Under these conditions, however, domain structures of different specimens with similar thicknesses can be compared to understand their different magnetic properties, as mentioned in Section 6.4. In addition, by application of an external magnetic field, detailed magnetization processes can be analyzed systematically by in situ electron holography and Lorentz microscopy, as discussed in the following paragraphs.

One of the nanocrystalline soft magnetic materials is $Fe_{73.5}Cu_1Nb_3Si_{13.5}B_9$, known as "Finemet." As reported elsewhere [13, 14], its lowest coercivity is obtained via an optimal heat treatment that results in mixing of the nanocrystalline phase with an amorphous phase. In such soft magnetic materials, the magnetization process can easily be achieved by tilting the specimen (Figure 8.4). Increasing the tilt angle enables the component of the external magnetic field or the stray field of the objective lens in the specimen plane to be gradually increased.

Figure 8.5 shows reconstructed phase images of $Fe_{73.5}Cu_1Nb_3Si_{13.5}B_9$ annealed at 823 K, where the tilt angle for the specimen was gradually increased. Because of the tilting, the component of the external magnetic field along the film plane gradually increased. As indicated in the figure, with increasing tilt angle, the lines of magnetic flux forming the closure domain changed gradually and eventually tended to be almost parallel because of the effect of the external magnetic field at a tilt angle of 4°. Note also that the leakage of lines of magnetic flux became pronounced with increasing tilt angle.

Soft ferrites (e.g. Mn–Zn ferrite, which is widely used in transformers) are another typical soft magnetic material [15, 16]. To understand their low coercivity, a magnetization process with an external magnetic field produced by an electromagnet (Figure 8.6a,b) was investigated.

Figure 8.7 shows the change in the reconstructed phase images of a Mn–Zn ferrite with increasing applied magnetic field [17]. When these reconstructed phase images were observed, the misalignment of the electronic optical axis and astigmatism were corrected. When the applied magnetic field was smaller than 1600 A m^{-1}, no change was observed in the distribution of the magnetic flux (Figure 8.7a,b). The distribution of the magnetic flux changed, however, from the state shown in Figure 8.7b to that shown in Figure 8.7c when the magnetic field was increased to 1760 A m^{-1} (refer to the domain-wall motion). In Figure 8.7d, the magnetic flux changes preferentially at the grain boundary, and a new domain wall, indicated by the black dashed line, appears at this boundary. This phenomenon appears to be due to the decrease in permeability at a grain boundary that includes precipitates such as SiO_2. However, a new domain wall indicated by a white dashed line appears in Figure 8.7f. In Figure 8.7h, this domain wall has moved to the right as a result of the increase in intensity of the applied magnetic field. Finally, in Figure 8.7i, the lines of magnetic flux have the same direction as that of the applied magnetic field on the right side of the image. By contrast, an antiparallel flux is observed on the left side. This flux was likely caused by a thick area formed during the focused ion beam (FIB) milling process.

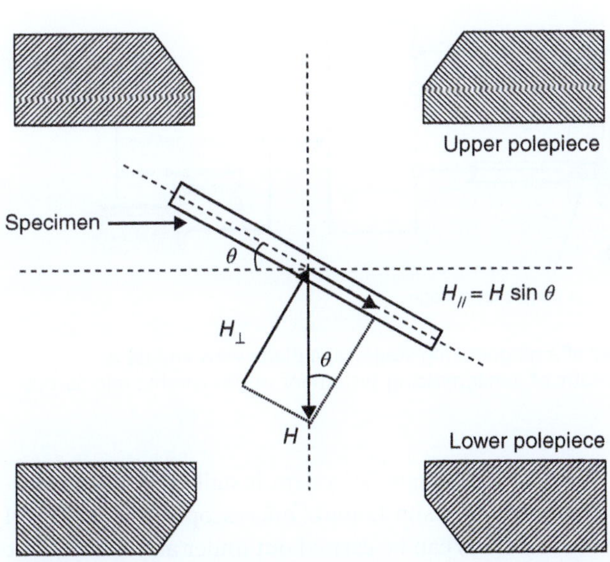

Figure 8.4 Illustration showing introduction of residual magnetic field of objective lens in film plane by tilting a specimen.

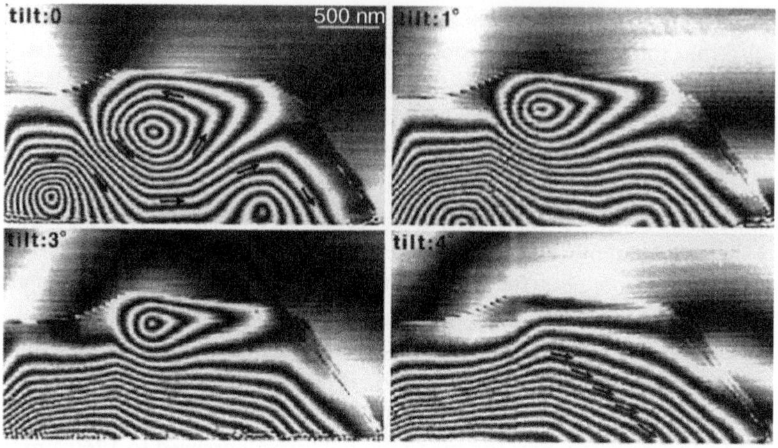

Figure 8.5 Change with increasing tilt angle in reconstructed phase images of $Fe_{73.5}Cu_1Nb_3Si_{13.5}B_9$ annealed at 823 K. Experiment was performed in presence of a residual magnetic field in pole piece, which differed from the magnetically shielded objective lens. In-plane components of external magnetic field are 0 (tilt 0°), 2.8 (1°), 8.3 (3°), and 11.1 A m^{-1} (4°). Source: Shindo et al. 2002, with permission from Elsevier.

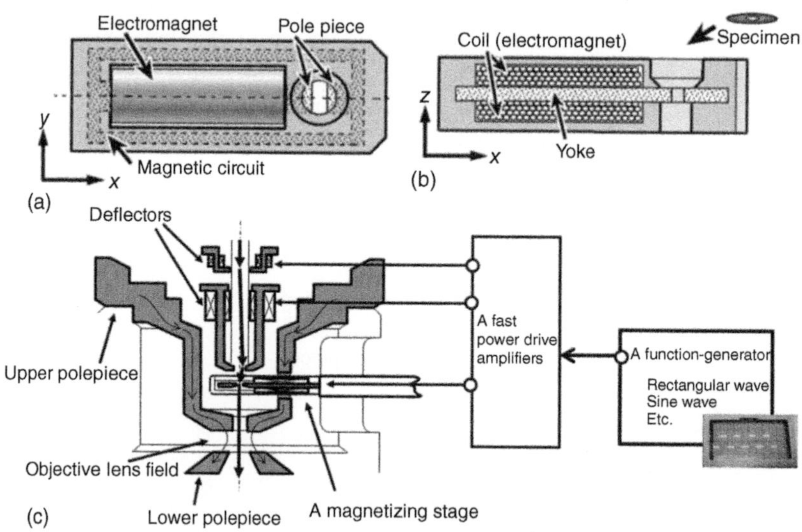

Figure 8.6 Schematic of head of a magnetizing stage: (a) a plane view and (b) a cross-sectional view. (c) Schematic of a magnetizing system for in situ Lorentz microscopy using an AC magnetic field.

By using an alternating current (AC) magnetic system installed on a transmission electron microscope (Figure 8.6c), in situ Lorentz microscope observations of domain walls in soft magnetic materials can be carried out under an AC magnetic field. In the aforementioned experiments (Figures 8.5 and 8.7), a DC magnetization system was used for electron holography. Because the exposure time for observing holograms is usually several seconds and interference fringes are sensitive to the fluctuation of the illumination angle due to the applied external magnetic field, dynamic observation of domain-wall motions by electron holography is difficult. For this purpose, Lorentz microscopy with Fresnel mode [18–23] is useful, as detailed in the following discussion. Grüendmayer and Zweck [24] used a thin metallic wire to apply an AC magnetic field. The use of a metallic wire was reported earlier by Yi et al. [20], although they achieved a pulse magnetic field rather than an AC magnetic field. However, Akase et al. used a small electromagnet to apply an AC magnetic field [22] to an electrical steel sheet [25] and $Fe_{85}Si_2B_8P_4Cu_1$ nanocrystalline soft magnetic alloy [26].

Figure 8.8 shows Lorentz micrographs (Fresnel mode) of a doubly oriented electrical steel sheet in an AC magnetic field; these micrographs were captured from a videotape [25]. The frequency and amplitude of the AC magnetic field were 0.34 Hz and 2.5 kA m^{-1}, respectively. The frames from (a) to (d) correspond to one-half of one AC cycle. The left column shows the original micrographs; in the micrographs on the right, the dashed lines and small arrows indicate the domain-wall contrast and magnetization, respectively. In addition, the arrowhead indicates a SiO_2 precipitate, and the large arrow at the top right indicates the direction of the applied magnetic field. From (a) to (d), we observe that the domain walls move smoothly; thus, the area of the domain whose magnetization is parallel to the applied magnetic field increases with increasing magnetic field. By contrast, the domain with magnetization antiparallel to the applied field disappears in (d), whereas the domains with perpendicular magnetization remain around the precipitate. Note that the domain walls tend to have a linear shape parallel to the axis of easy magnetization.

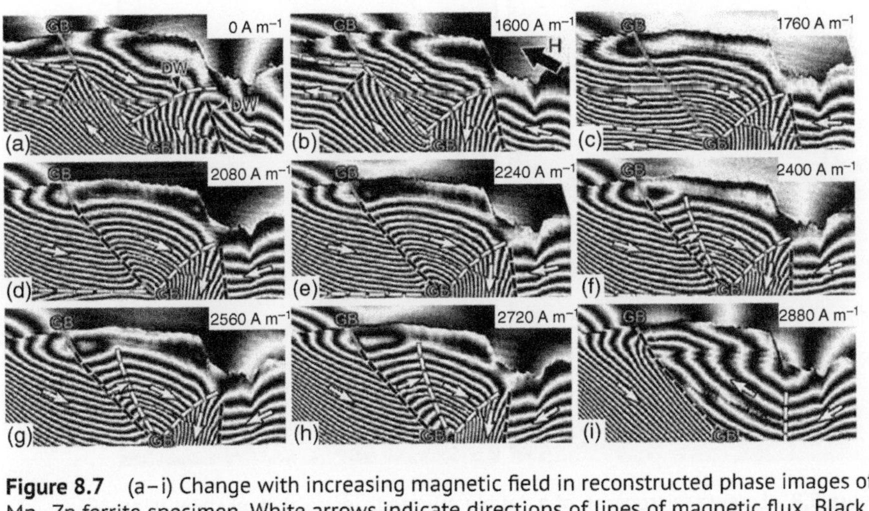

Figure 8.7 (a–i) Change with increasing magnetic field in reconstructed phase images of a Mn–Zn ferrite specimen. White arrows indicate directions of lines of magnetic flux. Black arrow in (b) represents direction of applied magnetic field. Broken lines indicate positions of magnetic domain walls (DWs). Line between "GB" indicates position of a grain boundary. Source: Shindo and Akase (2020) with permission from Elsevier.

Figure 8.9 shows Lorentz micrographs (Fresnel mode) of a non-oriented electrical steel sheet observed under the two-beam condition of a bright-field image [25]. Under this condition, the contrast of the strain fields around the precipitates and many dislocations is enhanced, thus clearly specifying the positions of these defects. When an AC magnetic field (sine wave, 1.0 Hz, 2.4 kA m^{-1}) is applied, the domain walls are trapped not only (a) at the precipitates but also (b) at the regions around their strain fields. As demonstrated here, dynamic observation of the interaction

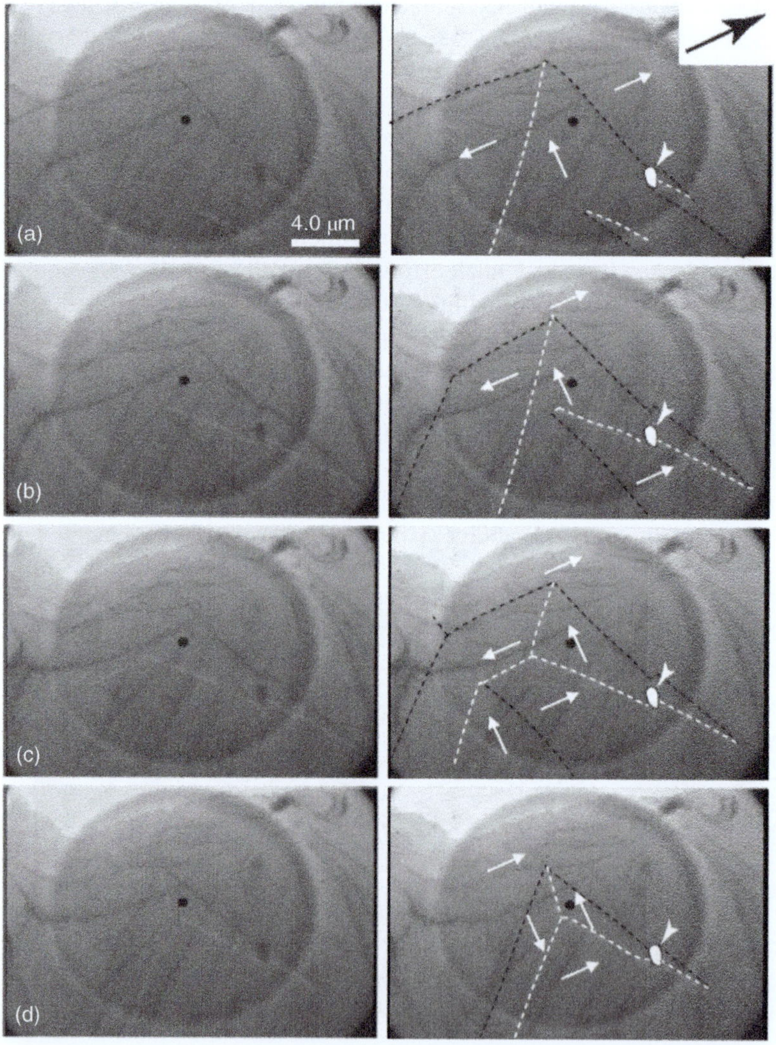

Figure 8.8 Lorentz micrograph of a doubly oriented electrical steel sheet in an AC magnetic field (0.34 Hz, 2.5 kA m^{-1}), as captured from a videotape. Large arrow at the top right indicates direction of applied magnetic field. In pictures on right, dashed lines and small arrows indicate domain-wall contrast and magnetization, respectively. Arrowheads indicate SiO_2 precipitates. Source: Shindo and Akase (2020) with permission from Elsevier.

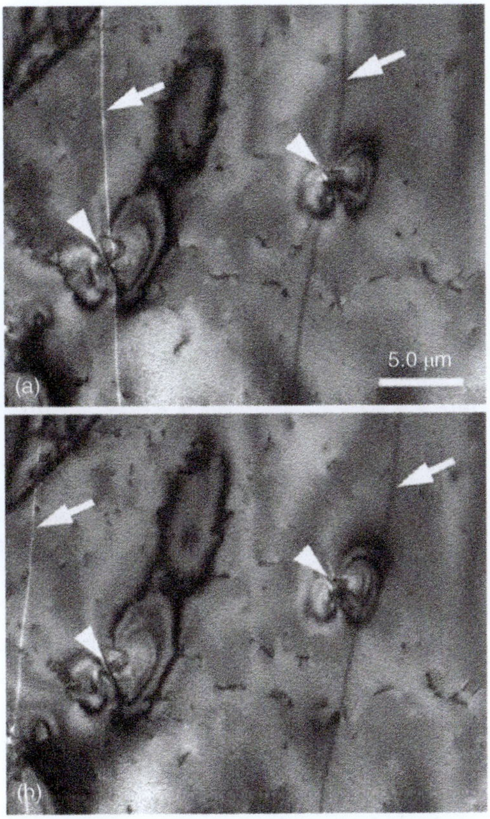

Figure 8.9 Lorentz micrographs of a non-oriented electrical steel sheet observed under two-beam condition of a bright-field image. Arrows and arrowheads indicate domain-wall contrasts and two AlN particles, respectively. Source: Shindo and Akase (2020) with permission from Elsevier.

between domain walls and lattice defects is important to understand eddy current and hysteresis losses in electrical steel sheets. Some of the inclusions have been introduced to improve the mechanical properties. In this experiment, domain walls are pinned at these inclusions and the strain field around them. The pinning process was found to depend on the size of the inclusions. Thus, this information can be used to improve not only mechanical properties but also magnetic properties.

8.2.2 Hard Magnetic Materials

Hard magnetic materials, sometimes called "permanent magnets," can be classified into two categories depending on the mechanism by which their large coercivity is achieved. The first mechanism is nucleation mechanism, and the second is pinning. The magnetization process of the former mechanism is governed by nucleation for magnetization reversal. However, for the latter mechanism, the coercivity results from the pinning process of domain walls. Notably, in thin specimens for TEM observation, the magnetic properties of hard magnetic materials differ dramatically from those of the corresponding bulk materials. Furthermore, unlike soft magnetic materials, hard magnetic materials are difficult to characterize by in situ electron holography using a large external magnetic field. Nevertheless, clarifying

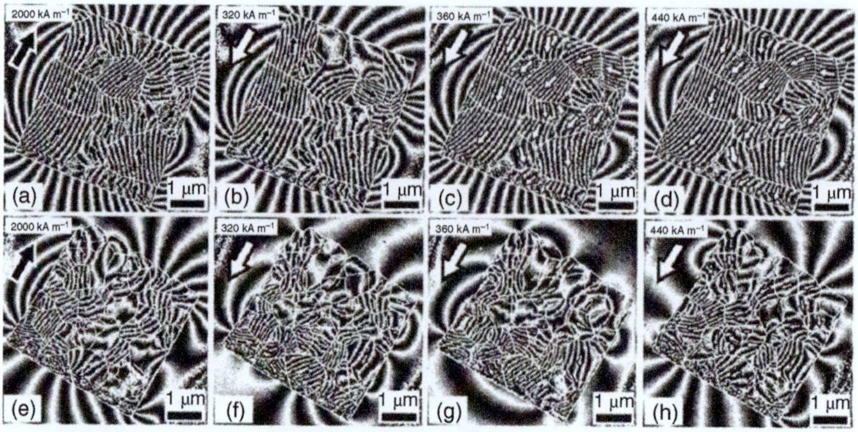

Figure 8.10 Reconstructed phase images of (a–d) anisotropic and (e–h) isotropic specimens of Ba ferrites observed in remanent states. White lines indicate specimen edges and grain boundaries. Large and small arrows indicate directions of applied magnetic field and the lines of magnetic flux, respectively. Source: Shindo and Akase (2020) with permission from Elsevier.

the detailed magnetic flux distributions by electron holography and comparing them with microstructures is important for understanding and improving the magnetic properties of hard magnetic materials.

A typical hard magnetic material with the nucleation mechanism is Nd–Fe–B, the main phase of which is $Nd_2Fe_{14}B$ [27, 28]. McCartney and Zhu [29, 30] used electron holography techniques to determine the domain-wall widths in die-upset $Nd_2Fe_{14}B$ magnets, with analysis of nanometer-sized domain walls. The change in the magnetic flux distribution at a domain-wall position was also investigated by Park et al. [31] for a $Nd_2Fe_{14}B$ magnet prepared by the melt-spinning method; the authors discussed the effect of the heat treatments in detail. Recently, the few-nanometer-wide boundary phases of a Ga-doped Nd–Fe–B sintered magnet were studied with high precision using the split-illumination holography technique described in Section 6.1. From detailed simulations of reconstructed phase images, the authors quantitatively analyzed the magnetization of the boundary phases [32].

Figure 8.10 shows the change in the remanent states of anisotropic and isotropic Ba ferrites [33], which were prepared with and without an applied magnetic field, respectively, before sintering. The large arrows indicate the direction of the applied external magnetic field, with the magnitude indicated at the top left of each image. The small arrows indicate the directions of the lines of the magnetic flux. Substantial magnetization reversal is observed in the anisotropic specimen (upper panels), whereas it is more gradual in the isotropic specimen (lower panels). This difference in the magnetization distribution during the magnetization process has been reported to correlate well with the demagnetization curves of these two specimens [34].

A typical hard magnetic material with the pinning mechanism is Sm–Co, which consists primarily of a Sm_2Co_{17} phase and a $SmCo_5$ phase. The coercivity of Sm–Co

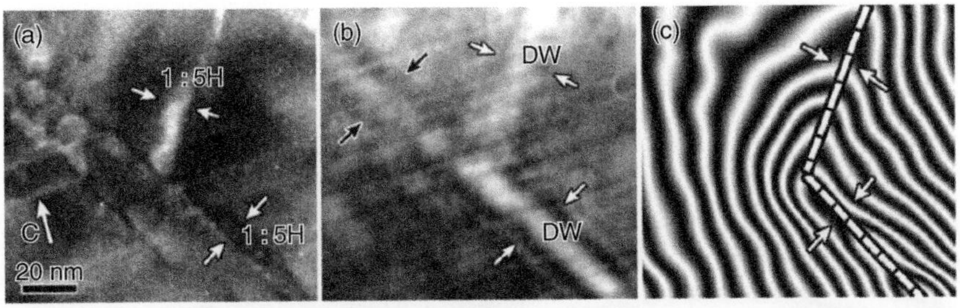

Figure 8.11 (a) Bright-field image, (b) Lorentz microscope image, and (c) reconstructed phase image observed in the same area of a hard magnetic material, $Sm(Co_{0.720}Fe_{0.200}Cu_{0.055}Zr_{0.025})_{7.5}$. Small white arrows indicate cell boundaries at which magnetic domain walls are present. Small black arrows in (b) represent cell boundaries at which magnetic domain walls are absent. Large white arrow shows direction of *c*-axis. Source: Shindo and Akase (2020) with permission from Elsevier.

increases upon the addition of Cu and upon an appropriate heat treatment. To understand the magnitude of the coercivity, clarifying the pinning positions of the domain walls is important. As shown in Figure 8.11 [35], which was obtained by coupling Lorentz microscopy in Fresnel mode with electron holography, the domain wall is located directly on the cell boundary phase (i.e. $SmCo_5$ containing Cu).

To investigate the magnetization process dynamically, it is necessary to apply a strong magnetic field to the specimen inside a transmission electron microscope. When an electromagnet placed in the specimen holder is used, the maximum magnetic field is limited to approximately 0.05 T. This limitation is due to the strong deflection of the incident electron beam because the magnetic field produced by the electromagnet is not confined to the specimen plane but spreads out above and below the specimen. However, a unique technique of producing a strong, localized magnetic field has been demonstrated [36], wherein a small, sharp magnetic needle of Nd–Fe–B was used to produce the magnetic field. As illustrated in Figure 8.12, the needle's position can be controlled using piezoelectric elements (Section 6.6) (i.e. the magnetic field can be increased by moving the needle to the specimen). Researchers have used this technique to observe the motion of the domain walls in a sintered $Nd_2Fe_{14}B$ specimen, as shown in Figure 8.13 [36]. In the figure, the

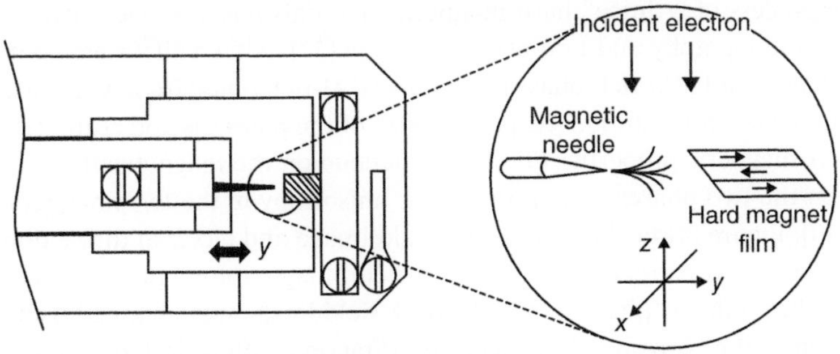

Figure 8.12 Schematic of part of a specimen holder equipped with a sharp magnetic needle.

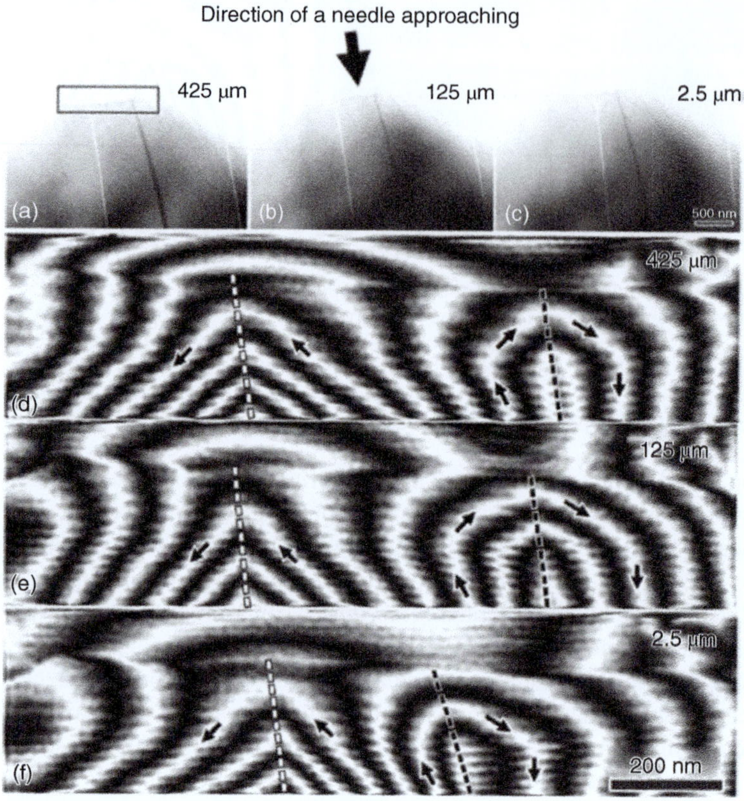

Figure 8.13 (a–c) Lorentz microscope images and (d–f) reconstructed phase images of sintered $Nd_2Fe_{14}B$. Reconstructed phase images correspond to rectangular region in (a). Each image shows distance between needle and specimen, with direction of needle's approach (large arrow) indicated at the top. Phase amplification was 2. Source: Shindo and Akase (2020) with permission from Elsevier.

change of domain structures in Lorentz microscope images (a–c) and corresponding reconstructed phase images (d–f) with the change of the distance between needle and specimen is presented.

The pinning process of "Alnico" hard magnetic materials has also been investigated by electron holography and Lorentz microscopy [37]. Alnico alloys are composed of two phases, an FeCo-rich phase (α_1) and a NiAl-rich phase (α_2), which are formed by spinodal decomposition. The particle size of the phases is approximately 20–40 nm. Their magnetic properties (Table 8.1) depend on the magnetization difference between the two phases and on the shape anisotropy in the ferromagnetic α_1 phase [38–40]. Figure 8.14a shows the dark-field image and electron diffraction pattern of demagnetized Alnico 8. In the bright-field image, the α_1 phase cannot be distinguished from the α_2 phase. Thus, the dark-field image was obtained using the 001 reflection in the corresponding electron diffraction pattern in Figure 8.14a because the 001 reflection at the [100] zone axis is only attributable to the α_1 phase [41]. Notably, the ferromagnetic α_1 and nonmagnetic α_2 phases appear as bright and

Table 8.1 Magnetic properties and compositions of Alnico 5 and Alnico 8.

					Composition (mass%) (balance Fe)			
No.	**Specimen**	$(BH)_{max}$ (kJ m^{-3})	B_r (T)	H_c (kA m^{-1})	**Al**	**Ni**	**Co**	**Cu**
1	Alnico 5	42.3	1.30	50.5	8.0	14.1	24.0	3.0
2	Alnico 8	44.8	0.91	116	7.3	14.1	35.5	3.0

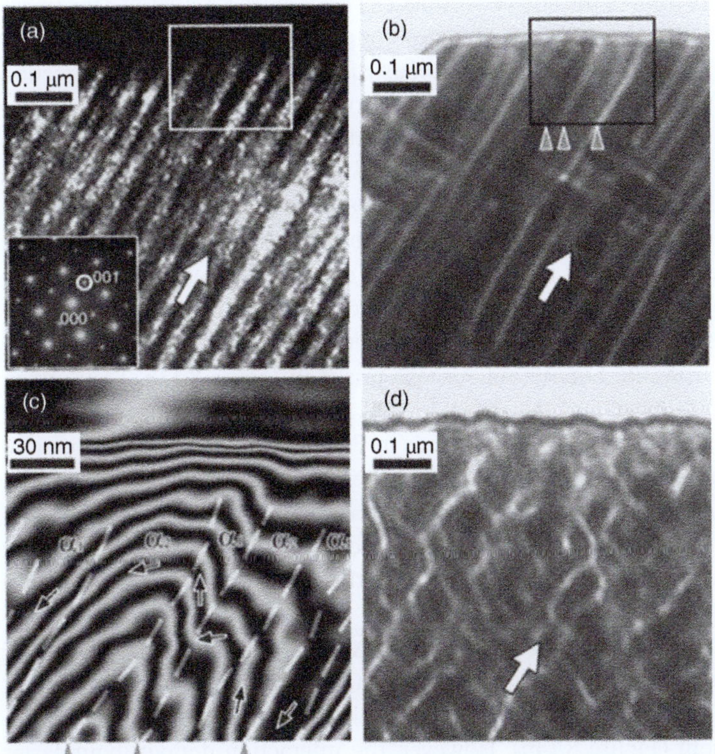

Figure 8.14 (a) Dark-field image and (b) Lorentz microscope image of demagnetized Alnico 8. (c) Reconstructed phase image obtained from region indicated by rectangle in (a) and (b). Broken white lines indicate the boundaries between α_1 and α_2 phases. (d) Lorentz microscope image of demagnetized Alnico 5. White arrow indicates direction of magnetic field applied in the thermomagnetic treatment, and black arrows indicate the direction of lines of magnetic flux.

dark regions, respectively. As shown in the dark-field image, the α_1 phases are well aligned in the direction of the magnetic field applied during the thermomagnetic treatment, as indicated by the white arrow, and have an elongated shape with a width of approximately 30 nm.

Figure 8.14b shows the Lorentz microscope image of demagnetized Alnico 8, which was observed using the Fresnel mode under a defocused condition. The white lines and black bands are aligned in the direction of the magnetic field applied during the thermomagnetic treatment. As explained in Section 6.3, in the Fresnel mode, domain walls are observed as white lines or black bands because the electron beam passing through the domains is deflected by the Lorentz force; thus, the electron intensity increases or decreases at the domain wall according to the magnetization direction of the adjacent domains [42]. Figure 8.14c shows the phase distribution and reconstructed phase image obtained from the region indicated by the rectangle in Figure 8.14a,b. In the reconstructed phase image, the direction and density of white lines, respectively, correspond to the direction and density of lines of magnetic flux projected along the electron beam. We assumed that the contribution of electrostatic potential could be ignored because the specimens were prepared using an FIB technique. In Figure 8.14c, the direction of the lines of magnetic flux tends to change at the boundaries between the α_1 and the α_2 phases as a result of the stray field of the ferromagnetic α_1 phase, which is thought to be a single domain. In addition, we note that the distribution of lines of magnetic flux in Figure 8.14c shows the detailed magnetic domain structure, including the information from the Lorentz microscope image in Figure 8.14b. Thus, when both the dark-field image and the reconstructed phase image are considered, the boundaries indicated by the gray arrowheads in Figure 8.14c clearly correspond to the white lines and the black band indicated by the gray arrowheads in Figure 8.14b. Therefore, the white lines and black bands observed by Lorentz microscopy can be considered the boundaries where the direction of the lines of magnetic flux changes at all the boundaries between the α_1 and the α_2 phases (interaction domain boundary [38, 43]).

A Lorentz microscope image of demagnetized Alnico 5 is shown in Figure 8.14d. The white lines and black bands, which indicate the interaction domain boundaries, have nearly the same direction as the magnetic field applied during the thermomagnetic treatment (Figure 8.14b). However, the interaction domain boundaries observed in Alnico 8 are almost straight (Figure 8.14d), whereas they appear as fluctuated line shapes in Alnico 5 (Figure 8.14d). Notably, Alnico 8 exhibits greater coercivity than Alnico 5 (Table 8.1), which is attributed to the large shape anisotropy of the α_1 phase. Therefore, the magnetic bamboo-like microstructure in Alnico 8 is reasonably considered to reflect the fact that Alnico 8 exhibits greater shape anisotropy than Alnico 5.

Figure 8.15a,b show enlarged Lorentz microscope images of Alnico 5, as captured from videotape. The time difference between Figure 8.15a,b is 0.066 seconds, and the distance between the specimen and magnetic needle, where the induced magnetic field was estimated by simulation to be approximately 56 kA m^{-1}, is ~50 μm.

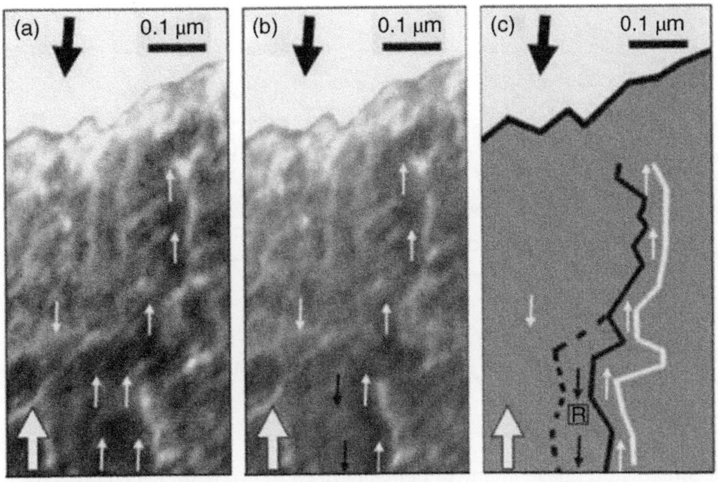

Figure 8.15 (a) and (b) Enlarged Lorentz microscope images of Alnico 5, as captured from a videotape. Distance between the specimen and magnetic needle is 50 μm, and time difference between (a) and (b) is 0.066 seconds. (c) Schematic showing magnetization process caused by magnetic field induced by magnetic needle. Large black and white arrows indicate direction of approach of magnetic needle and direction of magnetic field applied in thermomagnetic treatment, respectively. Small black and white arrows indicate magnetization direction; in particular, magnetization direction at magnetization-reversed domain is indicated by small black arrows.

Figure 8.15c is a schematic showing the magnetization process caused by the magnetic field induced by the magnetic needle. The interaction domain boundaries are indicated by the black and white lines. In Figure 8.15c, the magnetization direction of the region indicated by "R" was reversed; therefore, the boundary situated at the position indicated by broken black lines moved toward a new position shown by the solid black line. However, the distance between the solid black line and white line was evaluated to be approximately 40 nm [38–40]. Consequently, the magnetization is accomplished through the magnetization reversal of the α_1 phase. The distance between the solid black and white lines indicates that the mechanism of magnetization reversal can be regarded as a reversal of the entire magnetization in one grain of the α_1 phase through, presumably, the incoherent rotation mode [38, 39].

Figure 8.16a,b show the reconstructed phase images of Alnico 8 before and after a magnetic field of approximately 110 kA m^{-1} was induced by the magnetic needle. The reconstructed phase images were obtained after the magnetic needle was removed. Figure 8.16c,d show schematics of Figure 8.16a,b, respectively. After the magnetic field was induced, a change in the direction of lines of magnetic flux was observed at the regions indicated by "R." This result is attributed to the magnetization reversal of domains caused by the magnetic field induced by the magnetic needle, which is similar to the magnetization process in Alnico 5. In Figure 8.16b,d, the width of the domains, where the direction of lines of magnetic flux is opposite to the direction of the approaching needle (indicated by the small white arrows), is evaluated to be ~30 nm, corresponding to the width of the α_1 phase in Figure 8.14a.

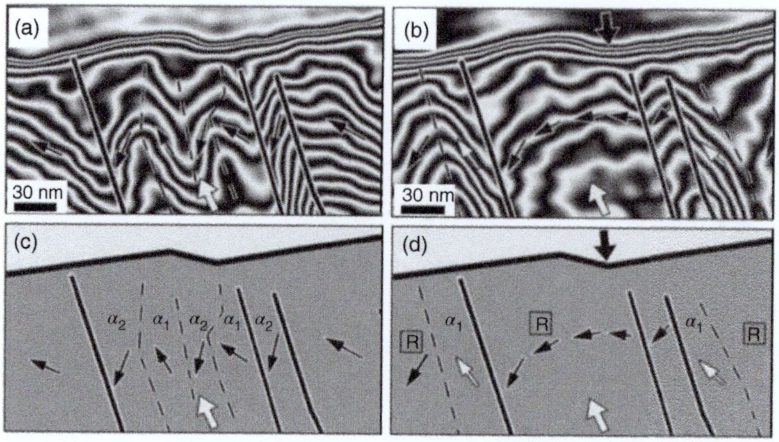

Figure 8.16 (a) Reconstructed phase image of demagnetized Alnico 8. (b) Reconstructed phase image after a magnetic field was induced using magnetic needle; (c) and (d) are schematics of (a) and (b), respectively. Solid and broken black lines indicate interaction domain boundaries. Interaction domain boundaries that are not changed when a magnetic field is induced are depicted by solid black lines. Small black and white arrows indicate direction of lines of magnetic flux. Large black and white arrows indicate direction of approach of magnetic needle and direction of magnetic field applied in thermomagnetic treatment, respectively.

Consequently, the magnetization reversal in each elongated grain of the α_1 phase is considered to be achieved in the incoherent rotation mode, as in the case of Alnico 5.

Figure 8.17 shows Lorentz microscope images of the nucleation process in $Nd_2Fe_{14}B$ for magnetization reversal [44]; these images were captured from videotape. The top-right corner in each image indicates the passage of time. The speed of the needle was estimated to be ~25 nm per frame (or 0.033 seconds). Using the electromagnet, we applied a magnetic field to the thin-foil specimen parallel to the direction indicated by the large blue arrow (H_{ext}) in Figure 8.17a and then removed it.

The magnetization direction in grains G1 and G2 was considered to be parallel to the axis of easy magnetization in each grain. However, in Figure 8.17b, the magnetized sharp needle approached the thin-foil specimen in the direction indicated by the large red arrow. Through simulation, the magnetic field at the top of the needle was estimated to be ~560 kA m^{-1}, whereas that at the specimen position in (a) was ~120 kA m^{-1}. Because the thin-foil specimen was being magnetized, no white lines or black bands, which correspond to magnetic domain walls, were observed until (a). However, a pair of domain walls abruptly appeared immediately after (a), as shown in Figure 8.17b. The domain walls, which appear as white lines and black bands, are shaped as straight lines and not parallel to each other. Notably, the nucleation of magnetization reversal occurs abruptly with a large volume of paired, reversed domains forming from and along the grain boundary. Because the magnetized needle approaches from the direction along the large red arrow and because the external

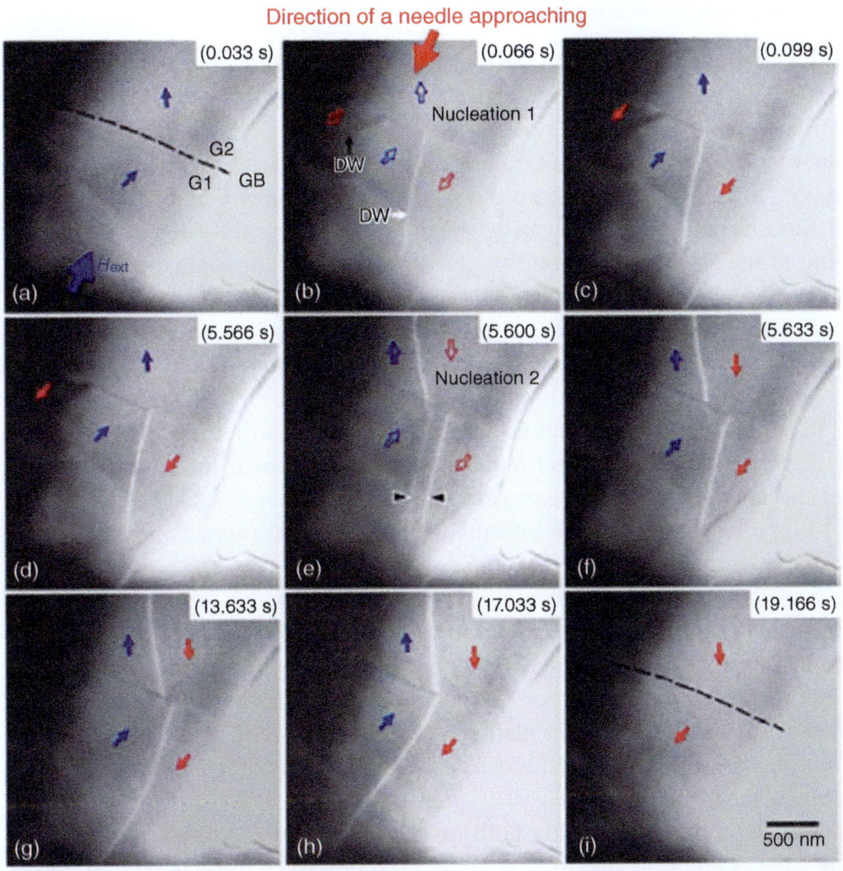

Figure 8.17 Lorentz microscope images captured from a videotape, showing nucleation process for magnetization reversal in $Nd_2Fe_{14}B$. (a) Remanent state. (b) Reversed domains form in grain G1 from grain boundary. (c and d) Magnetic domain walls change gradually. (e) When one of domain walls disappears in grain G1, a domain wall appears in grain G2. (f and g) Domain walls in grains G1 and G2 move continuously. (h) Domain walls become connected at grain boundary and (i) move to left until they disappear. Passage of time is indicated in parentheses. GB and DW represent grain boundaries and domain walls, respectively. Blue and red arrows indicate direction of magnetization in each domain. Open arrows indicate magnetization direction at initiation of nucleation. Source: Park et al. [44], with permission from AIP Publishing LLC.

magnetic field is larger in grain G2 than that in grain G1, the nucleation of magnetization reversal is expected to occur first in grain G2. Interestingly, however, a pair of magnetic domain walls is formed preferentially not in grain G2 but in grain G1 (Figure 8.17b). This result indicates that the nucleation of magnetization reversal strongly depends on the shapes and properties of the grain boundaries. Thus, controlling the shapes and properties of the grain boundaries is important to improve the magnetic properties (e.g. the coercivity) of these hard magnetic materials. After the reversed domains are formed, the magnetic domain walls move continuously (for details, see [44]).

8.2.3 Magnetic Recording Materials

Several electron holography studies have been carried out to examine patterned magnetic dots, and these studies have provided useful information for the development of magnetic recording systems. Dunin-Borkowski et al. [45] manufactured submicrometer-sized patterned Co elements and applied a magnetic field to them by exciting the objective lens of a transmission electron microscope. They substantiated the magnetic interaction between the neighboring Co elements via direct imaging of the lines of magnetic flux. Heumann et al. [46] studied the switching behavior of small magnetic circular dots composed of permalloy. They reported that the dots (diameter: ~150 nm; thickness: 6 nm) remained in a single-domain state during the entire switching process, whereas magnetization reversal occurred for larger or thicker dots via the formation of a C-shaped state or an intermediate vortex state. In addition to these studies for patterned magnetic structures, Park et al. [47] used the electron holography technique to reveal the magnetic flux distribution in a model specimen of magnetoresistive random access memory (MRAM) bits. Kasama et al. [48] provided quantitative information about the magnetic remanent state in pseudo-spin-valve bits.

We considered a Co–CoO obliquely evaporated tape, which is a classical recording material that has been widely used in commercial applications. To improve the recording density, sputtered magnetic tapes have recently been intensively studied [49]. For magnetic recording tapes, Osakabe et al. [50] reported a pioneering holography study in which they used a plan-view Co film specimen. They also studied the distribution of stray magnetic fields of a perpendicularly recorded magnetic tape and a bubble memory device [51].

The magnetic flux distribution both inside and outside a cross-sectional specimen of Co–CoO tape, as shown in Figure 6.33, is described in Section 6.7. Xia et al. [52] carried out intensive computer simulations of the remanent state of a recorded Co–CoO tape. Figure 8.18 shows a simulated reconstructed phase image showing the lines of magnetic flux both outside and inside the specimen. The result agrees well with the observed results shown in Figure 6.33. A benefit of the simulation is that it reveals the magnetization distribution in the field of view. Figure 8.19 shows the magnetization vectors inside the tape, which are derived from the portion including the magnetic flux loops A and B in Figure 8.18. The length of the arrow is proportional to the mean magnetization in each portion. Figure 8.20 shows the change in the magnetic flux distribution in the remanent states after an external magnetic field as strong as 0.30 T was applied.

Kim et al. [53] observed the magnetic domains in a Co–Ni–Fe writer pole tip for a perpendicular recording head. In addition to the magnetization process of the writer pole tip, Figure 8.21 visualizes the small stray field around the tip in the remanent state. The magnetic flux is closed in Figure 8.21a. In the Co–Ni–Fe pole tip (Figure 8.21b), the first motion of the domain walls is observed when the applied field is ~4.0 kA m^{-1}. With an increase in the applied magnetic field (Figure 8.21b–d), the centers of the closure domains shift toward the opposite direction and disappear at the side edges of the pole tip, where the domain walls are pinned. The lines of

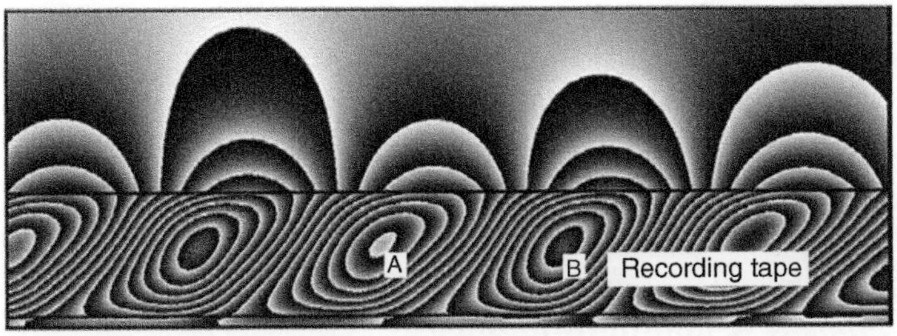

Figure 8.18 Simulated reconstructed phase image of a recorded Co–CoO tape (remanent state). Phase is represented in terms of ϕ rather than $\cos\phi$.

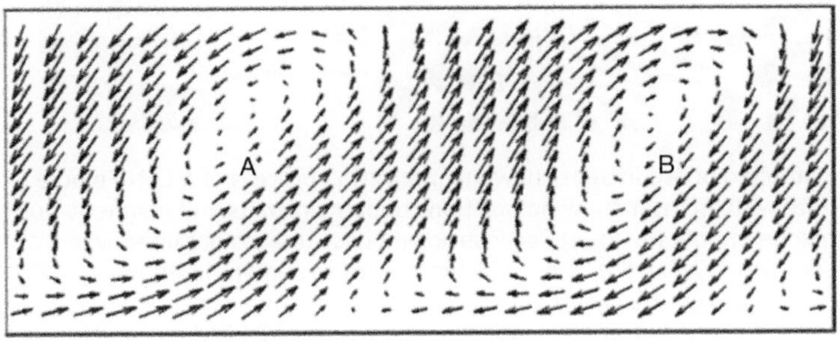

Figure 8.19 Magnetization distribution inside the recorded tape. This area corresponds to that containing the magnetic flux loops at A and B in Figure 8.18.

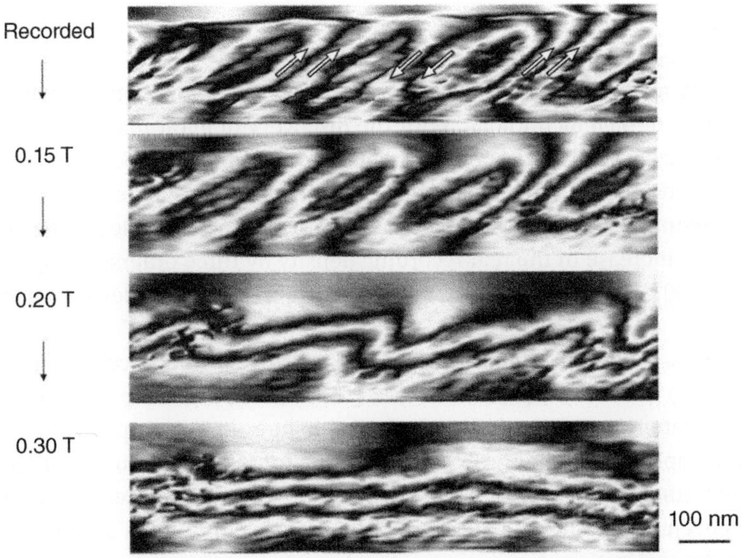

Figure 8.20 Change of magnetic flux distribution in remanent states of a recorded Co–CoO tape observed as a function of applied external magnetic field (from 0.15 to 0.30 T). Source: Shindo and Akase (2020) with permission from Elsevier.

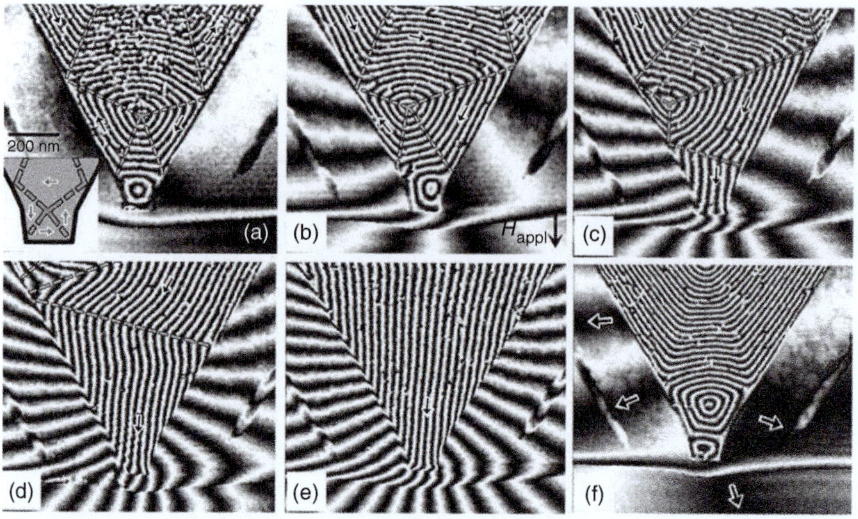

Figure 8.21 Reconstructed phase images show magnetization process of a CoNiFe pole tip. (a) Remanent state with an inset showing positions of domain walls and magnetic flux direction at bottom edge of pole tip. (b–e) Reconstructed phase images obtained when applied magnetic field was 4.0, 6.4, 8.0, and 12.0 kA m^{-1}, with specimen tilt angles of 4.8°, 7.7°, 9.6°, and 14.5°, respectively. Black arrow in (b) indicates direction of applied magnetic field. (f) Remanent state obtained after applying a magnetic field of 24.0 kA m^{-1} in the direction opposite that indicated by black arrow in (b). Source: Kim et al. [53], with permission from AIP Publishing LLC.

magnetic flux move in the direction of the applied magnetic field. Finally, the pole tip is fully magnetized (Figure 8.21e).

Ultrahigh density recording in hard disk drives (HDDs) is a production challenge for future advanced storage technology. With the discovery of giant magnetoresistance (GMR) [54], read heads and magnetic sensors with improved performance have been developed for HDDs; however, these read heads and magnetic sensors also require more complicated fabrication methods, with nanometer-scale precision. This work led to an increase in the real recording density by three orders of magnitude (from 0.1 to 100 Gbit $in.^{-2}$) and the emergence of spin-electronic devices such as MRAM [55].

Numerous factors, however, remain unclear and negatively influence the magnetoresistance (MR) effect and spin electronics [56, 57]. They include interfacial and barrier spin scattering, magnetic flux leakage, changes in the output voltage, and small signal amplitudes. Other problems arise from increasing the miniaturization from the micrometer to the nanometer scale in spin-electronic devices. Electron holography has been used to investigate some of these problems, which are mainly associated with magnetization. For example, the micromagnetic structure in individual MRAM bits has been observed by electron holography [47]. The magnetic field generated by the magnetic recording head in an HDD was also visualized by in situ electron holography [58]. In addition, the magnetization in a 5-nm free layer

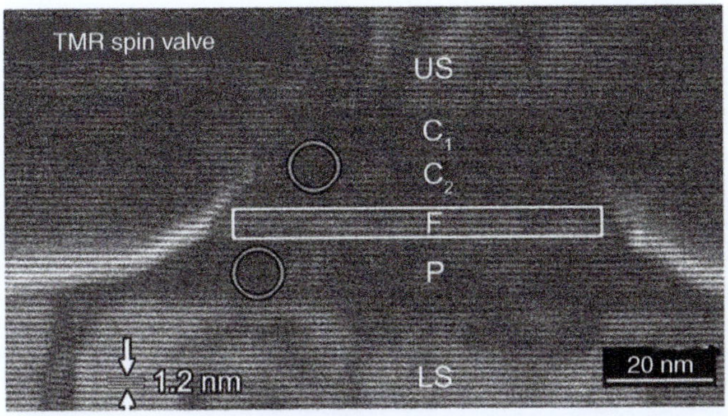

Figure 8.22 Hologram of a TMR spin-valve head, which was cut from original hologram (4096 × 4096 pixels). Labels in image represent each layer according to its functional properties: US (upper shield), C_1 and C_2 (cap), F (free), P (pinned), and LS (lower shield). The figure indicates fringe spacing of 1.2 nm.

in a tunneling magnetoresistance (TMR) spin-valve head has been observed [59], as discussed in the next paragraph.

The hologram in Figure 8.22 shows the complex hetero-nanostructures of the spin-valve head, which consists of various materials with magnetic or nonmagnetic properties. Depending on their functional properties, these structures are designated as cap (C_1, C_2), upper shield (US), lower shield (LS), free (F), and pinned (P) layers; thus, this specimen contains numerous interfaces.

Figure 8.23 shows the magnetization distribution in the remanent state, as obtained by averaging phase images reconstructed from 10 holograms acquired consecutively under the same conditions, which provides improved phase resolution. In the magnetic flux map, the direction and density of the black (or white) contour lines correspond to those of the in-plane component of the magnetic flux density projected along the electron beam. Thick contour lines are clearly observed in both the NiFe (not colored) and CoPt (blue) layers. In addition, the contour line is visible inside the 5-nm NiFe (free layer) below the dotted line in the yellow region. In the CoFe layer (red), the in-plane component of the magnetic flux appears to be zero (or extremely small), meaning that its magnetization direction is normal to the film plane. Figure 8.24 shows the change in the magnetization distribution under an applied field of 14 kOe normal to the film plane [59]. The thick contour lines in the B and LS layers (circled in yellow) disappear at 14 kOe (Figure 8.24a,b). This effect was reproduced in the phase-shift plot (φ_M) for the LS layer (Figure 8.24c), which indicates that the materials were fully magnetized parallel to the direction of the applied field.

Topological spin textures have been attracting increasing interest in the quantum magneto-transport field because of their possible applications in spintronics. Skyrmions are particularly attractive for use as information carriers in memory and logic devices because of the emergence of spin-transfer torque at extremely low current densities (1×10^6 A m^{-2}) [60]. A skyrmion carrying a topological quantum number acts as an effective magnetic flux. Charge carriers flowing over the

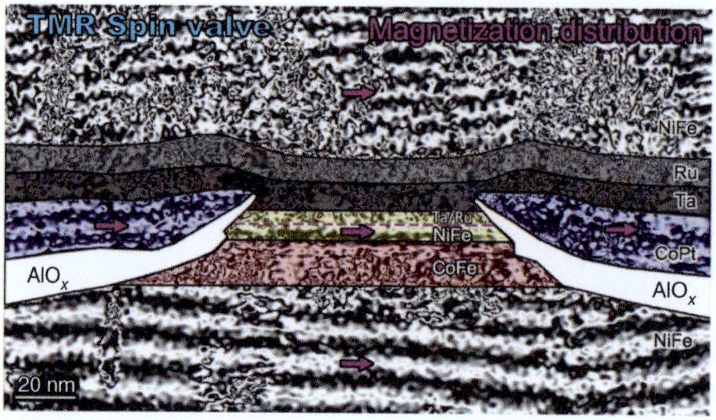

Figure 8.23 Magnetization distribution of a TMR spin-valve head. The reconstructed phase image was amplified by 6. The noisy and relatively thinner lines in the image are attributed to diffraction contrast and phase amplification. The arrows represent directions of the in-plane component of magnetization.

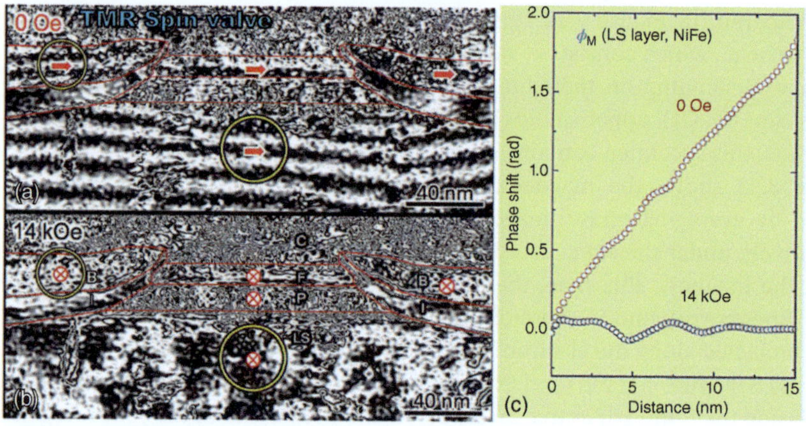

Figure 8.24 Change in magnetization distribution with applied field. (a) Reconstructed phase image under no applied field. (b) Reconstructed phase image obtained with a 14 kOe field applied perpendicular to film (image) plane. Phase amplification is 6. (c) Plot of phase shift ϕ_M in LS layer at 0 Oe and 14 kOe.

skyrmion crystal are deflected by the emergent electromagnetic fields induced by this magnetic flux, giving rise to the topological Hall effect [61]. Several challenges must be overcome before skyrmions can be used in actual devices. These challenges include clarifying their three-dimensional structures and fabricating thin films that contain skyrmions. Despite recent theoretical studies, the three-dimensional structures of skyrmions remain elusive. The observation of the three-dimensional structures of skyrmions at the microscopic level is a prerequisite for their application in spin-electronic devices. Real-space imaging of skyrmion lattices in thin specimens of semiconductors and insulators has been carried out using the Fresnel method of Lorentz microscopy [62, 63]. Quantitative analysis of the magnetic flux flow inside and outside a skyrmion is difficult, however, because of both the resolution limit (due to the defocus condition) and unwanted artifacts (surface roughness or contamination) of thin specimens.

Here, the two-dimensional magnetic flux distributions of a skyrmion lattice in a thin specimen of the helimagnet $Fe_{0.5}Co_{0.5}Si$ were investigated, and the three-dimensional structures of the helical and skyrmion phases were estimated [64]. As shown in Figure 8.25, the thickness of the specimen varied. The upper image in Figure 8.26a shows the two-dimensional phase shift of a skyrmion lattice, and the lower image shows the phase map represented by $\cos\phi(x,y)$. The black and white contour lines represent the lines of the magnetic flux density projected in the electron beam direction, and the red arrows indicate the direction of the in-plane component of the magnetic flux. An anticlockwise magnetic flux flow is observed in each skyrmion. The phase difference between the two black contour lines corresponds to $\Delta\phi = \pi/25$ rad [64].

The three-dimensional structures of the helical and skyrmion phases along the z-axis are shown in the lower-left corner of Figure 8.26b. The phase shifts for different specimen thicknesses as large as 500 nm were measured using the high penetration power of a 1 MV holography electron microscope. As shown in the graph, the shifts were linearly dependent on the specimen thickness, with a slope of 0.00173 rad nm^{-1} for both a helix (red triangles) and a skyrmion (black circles). The phase-shift characteristics of the skyrmion were similar to those of the helix.

The right side of Figure 8.26b shows a schematic of the three-dimensional spin configuration in a skyrmion and its electron phase shift along the A–B line (in the x–z plane), integrated over specimen thickness t. The colored arrows represent the magnetization direction at each point. We emphasize that, experimentally, the phase shift following the A–B line shows not only a sinusoidal shape but also a linear function with a stepped thickness from 55 to 510 nm (i.e. $\varphi = 0.00173\,t$ rad). These electron holography results indicate that the three-dimensional spin configuration of a skyrmion in the z-direction is cylindrical, similar to that of an $Fe_{0.5}Co_{0.5}Si$ helimagnet.

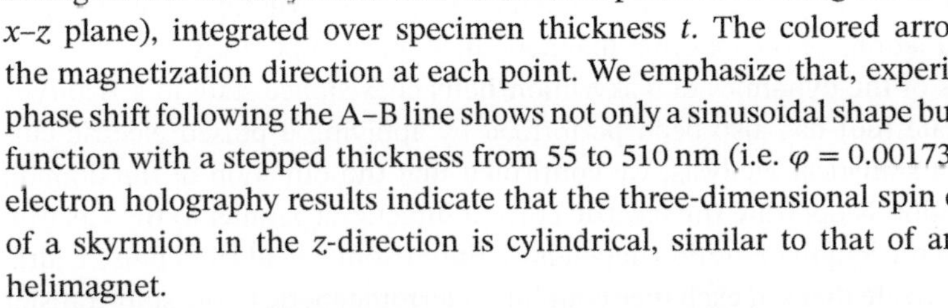

Figure 8.25 Image and schematic of a thin $Fe_{0.5}Co_{0.5}Si$ specimen produced using a focused ion beam. Thickness differences are represented as different levels of contrast. H_{ext} represents applied field. Inset on the left is diffraction pattern for electrons incident along [001] zone axis.

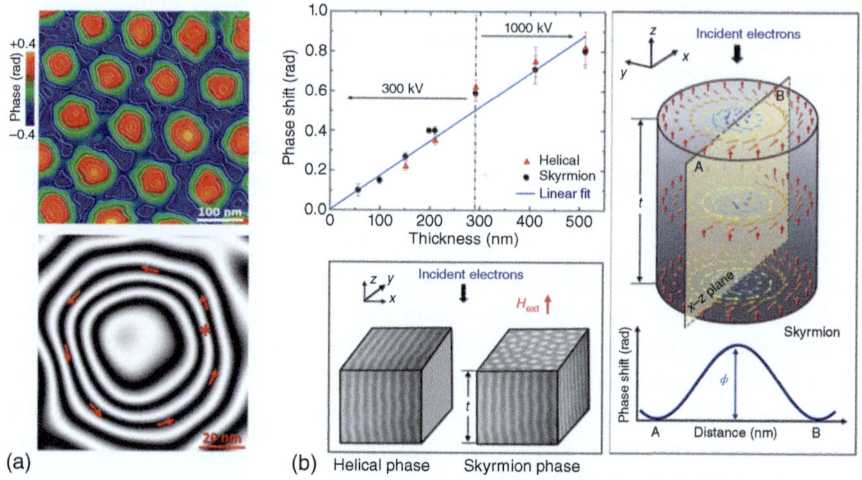

Figure 8.26 Two-dimensional phase maps and three-dimensional structure of the skyrmion lattices in a thin $Fe_{0.5}Co_{0.5}Si$ specimen. (a) Magnetic flux maps showing phase shift due to magnetic vector potential at $H_{ext} = 25$ mT and $T = 10.6$ K. Difference between contour lines corresponds to 0.1 rad. Black and white lines correspond to lines of magnetic flux. (b) Electron phase shifts of helix and skyrmion as a function of the specimen thickness (obtained using 300 kV and 1 MV microscopes) and a schematic showing three-dimensional spin structure of skyrmion and its phase profile for line A–B (in the x–z plane), integrated over specimen thickness t.

Magnetic skyrmions stabilized in $Fe_{1-x}Co_xSi$ nanowires have also recently been studied by electron holography, where detailed magnetic structure changes were clarified as functions of the external magnetic field and temperature [65]. Lorentz TEM analysis of the dynamics of a skyrmion/helix coexistence state in a B20-type FeGe thin plate [66] has also been performed by applying a pulsed electric current. Through statistical analysis, we confirmed that the direction of the domain boundary motion is opposite the electric current direction (parallel to the electron flow) and that the length of motion depends on the current density and pulse width. The results indicate that *s–d* exchange coupling is ferromagnetic in the spin-transfer torque mechanism. The boundary of the skyrmion lattice and helical domains has also been found to be more mobile at higher temperatures, which indicates that thermal agitation from the pinning potential affects the dynamics.

8.2.4 Ferromagnetic Shape-Memory Materials

Ferromagnetic shape-memory alloys (SMAs) provide a large recoverable strain, which is achieved through rearrangement of crystallographic domains in a martensitic phase (low-temperature phase) under an applied stress. In ferromagnetic SMAs (e.g. Ni_2MnGa [67, 68], Fe–Pd [69], Fe_3Pt [70], Co–Ni–Al [71, 72], and Ni–Ga–Fe [73]), shape deformation can be achieved by an applied magnetic field and by stress. As proposed by Ullakko et al. [67], the high magnetocrystalline anisotropy of martensites, which is typically on the order of 10^5 J m^{-3}, appears to be key in explaining the magnetic field-driven shape deformation. Regarding the magnetic domain structure, the theory of Ullakko et al. predicts a one-to-one correspondence between the magnetic domains and the crystallographic domains (martensite variants). Thus, to understand the peculiar shape deformation in ferromagnetic SMAs, revealing the magnetic domain structure is particularly important. Magnetic domains in martensites have been observed by Lorentz microscopy [74, 75], magnetic force microscopy (MFM) [76], the interference-contrast colloid technique [77], and similar methods. Notably, Venkateswaran et al. [78, 79] clarified the magnetic flux distribution in Ni–Mn–Ga alloys using Lorentz microscopy. They solved the transport-of-intensity equation (TIE), i.e. Eq. (6.8) with a focus series of Lorentz micrographs. Many ferromagnetic SMAs contain antiphase boundaries (APBs), which are structural defects in ordered alloys. In fact, APBs affect the magnetization distribution (especially in the parent phase) in ferromagnetic SMAs, as demonstrated by Murakami et al. [80, 81] and Venkateswaran et al. [78].

For example, Figure 8.27a provides a Lorentz microscope image of the parent phase of a $Ni_{50}Mn_{25}Al_{12.5}Ga_{12.5}$ alloy [82]. The magnetic domain walls are located at the positions of APBs. They are therefore heavily curved in the viewing field. These results explicitly indicate that the magnetic domain walls are stabilized at the positions of APBs. That is, APBs provide substantial pinning sites for domain wall motion, and Yano et al. [83] verified this prediction, as shown by the observed results in Figure 8.28. In the presence of these APBs, the lines of magnetic flux take the form of vortices (arrowheads in Figure 8.27b). For ease of comparison, the reconstructed phase image is superimposed on the Lorentz micrograph in Figure 8.27c. Another study [82] revealed that the effect of APBs is still prominent in the martensitic phase in a $Ni_{53}Mn_{25}Al_{11}Ga_{11}$ alloy.

8.3 Observation of Magnetic Structure Change with Temperature

Colossal magnetoresistance (CMR) in perovskite-type manganites has attracted considerable attention [84, 85]. A number of researchers have devoted time to clarifying the mechanism of CMR in manganites. In particular, several researchers have reported the importance of "phase separation," in which magnetic and nonmagnetic phases coexist, in the CMR effect [86–88]. This model highlights the magnetic microstructure. For example, assuming that ferromagnetic (FM) islands with metallic properties, which are formed in the nonmagnetic insulator matrix, are percolatively connected by applied magnetic fields, this connection gives rise to a large conduction path, thereby reducing the resistivity.

Several microscopy studies have been carried out to demonstrate the occurrence of magnetic phase separation and/or a mixed-phase state in hole-doped manganites. Uehara et al. [88] conducted Lorentz microscopy observations of a ceramic specimen $La_{5/8-y}Pr_yCa_{3/8}MnO_3$ ($y = 0.375$), wherein a submicrometer-scale mixture of "FM metal" and "non-magnetic insulator" phases was discovered at temperatures below the Curie temperature.

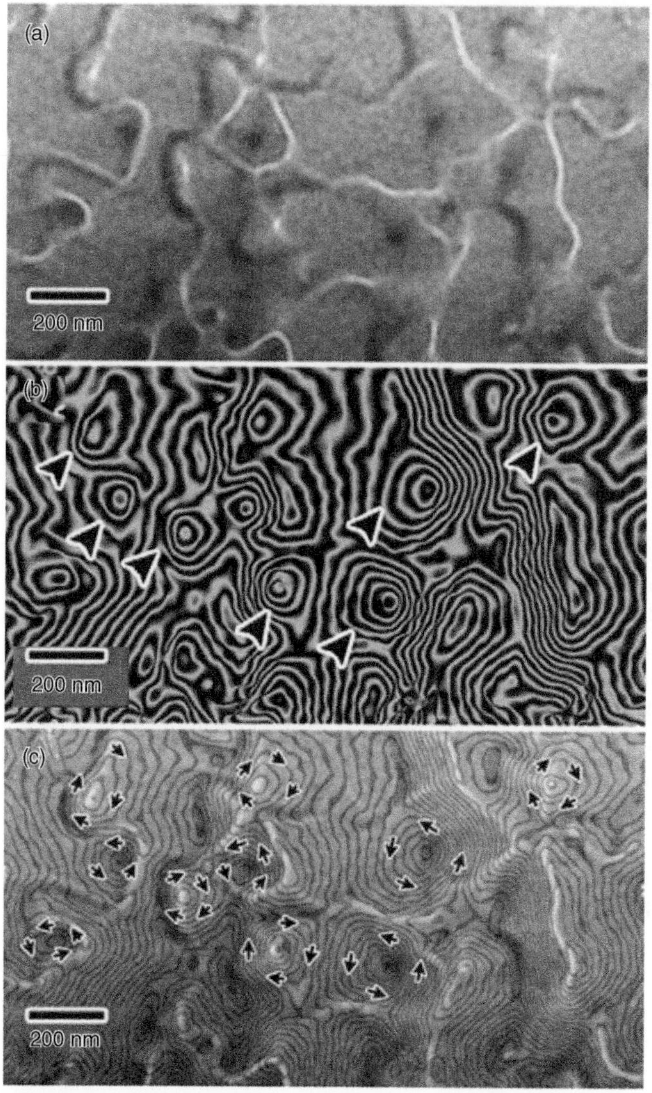

Figure 8.27 (a) Lorentz microscope image and (b) reconstructed phase image representing lines of magnetic flux in parent phase of $Ni_{50}Mn_{25}Al_{12.5}Ga_{12.5}$. Arrowheads show positions of magnetic vortices. (c) Superimposed images of (a) and (b), wherein small arrows indicate direction of magnetic flux. Source: Shindo and Akase (2020), with permission from Elsevier.

On the basis of magneto-optical imaging of a bulk specimen, Tokunaga et al. [89] reported percolative connections of separated FM domains in $(La_{1-y}Pr_y)_{0.7}Ca_{0.3}MnO_3$ ($y = 0.7$), which were induced by an electric current. They reported the effect of current-induced local heating to account for their observations. Zhang et al. [90] used the MFM technique to investigate the temperature dependence of both the microstructure and volume fraction of the FM phase

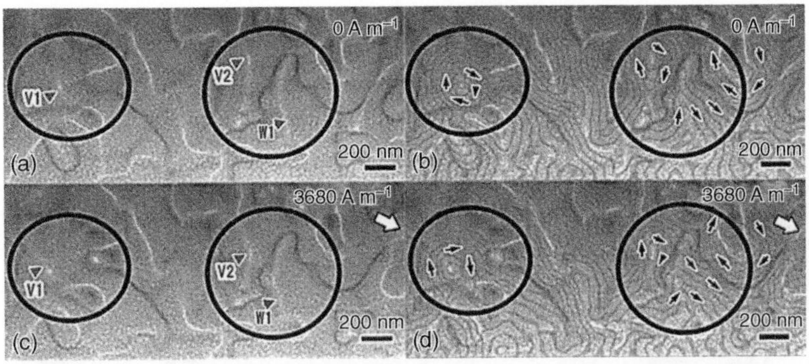

Figure 8.28 Magnetization process of a $Ni_{50}Mn_{25}Al_{12.5}Ga_{12.5}$ alloy in parent phase. (a and c) Changes in Lorentz microscope images with applied magnetic field. (b and d) Changes in reconstructed phase images with applied magnetic field. Reconstructed phase images are superimposed on Lorentz microscope images. White arrows indicate direction of applied magnetic fields.

in $La_{0.33}Pr_{0.34}Ca_{0.33}MnO_3$. In addition to these studies, other studies have observed magnetic domains in manganese oxides by Kerr microscopy [91], scanning Hall microscopy [92], and Lorentz microscopy [93].

In the case of electron holography, Loudon et al. [94] discovered a new phase (charge-ordered FM phase) in $La_{0.5}Ca_{0.5}MnO_3$, although their study focused on characterizing this new phase rather than analyzing the microstructure in the mixed-phase state. Regarding the mixed-phase state, another study examined the manganite $La_{0.44}Sr_{0.56}MnO_3$, which undergoes successive magnetic phase transformations (paramagnetic [PM] → AFM → FM) upon cooling [95]. Figure 8.29 shows reconstructed phase images corresponding to temperatures near the Néel temperature [96], where the phase information due to the mean inner potential has been removed. The phase images show that a region of the FM phase (indicated by an arrowhead) is isolated in the AFM matrix. The volume of the FM phase increases monotonically upon heating, via the motion of the phase boundary. Moreover, in the absence of an applied magnetic field, the magnetic flux is closed within the FM region at any stage of the phase separation. This result implies that

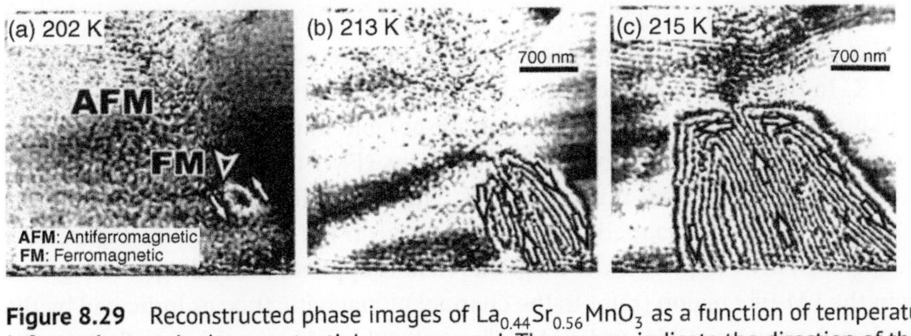

Figure 8.29 Reconstructed phase images of $La_{0.44}Sr_{0.56}MnO_3$ as a function of temperature. Information on the inner potential was removed. The arrows indicate the direction of the lines of magnetic flux.

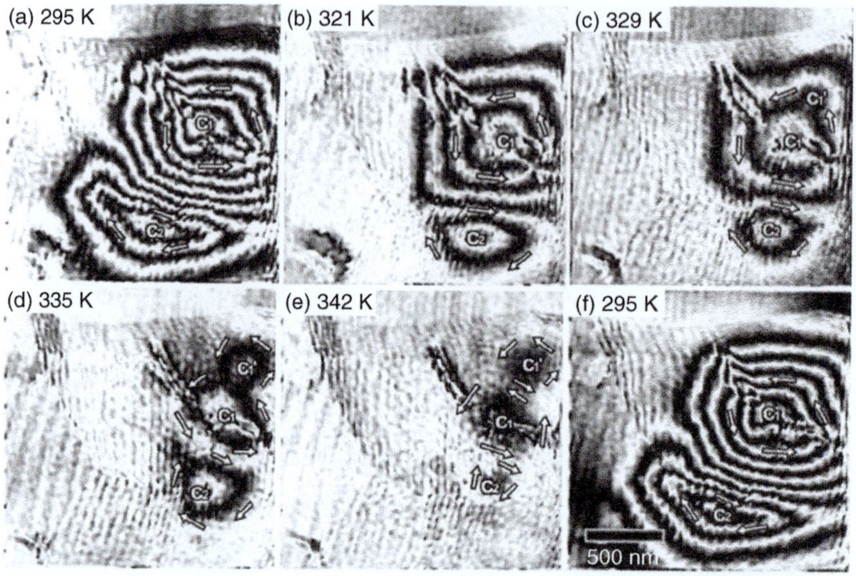

Figure 8.30 Reconstructed phase images of $La_{0.46}Sr_{0.54}MnO_3$ as a function of temperature. Information related to inner potential was removed. Arrows indicate direction of lines of magnetic flux. Source: Yoo et al. [97], with permission from American Physical Society.

magnetic interaction is negligible between well-separated FM islands under no applied magnetic field.

Another type of magnetic phase transformation, the FM → PM transformation, was observed for the manganite $La_{0.46}Sr_{0.54}MnO_3$ (Figure 8.30) [97]. Because the FM → PM transformation is similar to a second-order transformation, the spontaneous magnetization is gradually reduced (note the coarsening of the contour lines upon heating). Interestingly, the closure domain (C_1, Figure 8.30a) splits into two parts (C_1 and C_1', Figure 8.30e), both of which exhibit the same helicity as the magnetic flux.

In addition to studies of the change in the magnetic structure of perovskite-type manganites with temperature, studies of the magnetic microstructure near the phase-transformation temperature using an applied external magnetic field have also been reported (Figure 8.31) [98]. The specimen was $La_{0.81}Sr_{0.19}MnO_3$, which provides an inhomogeneous magnetic microstructure near the Curie temperature of 302 K. In this specimen, the FM regions (hatched portions in Figure 8.31a) and PM-like regions (white portions with no observable magnetization) are observed to coexist at room temperature. Two FM regions (FM1 and FM2) are observed in the absence of an external magnetic field. The magnetic flux is closed inside the FM regions; thus, the magnetic interaction between FM1 and FM2 in Figure 8.31 is negligible. However, when the magnetic field is applied, the leakage of magnetic flux in the PM-like region (refer to the channel of magnetic flux, as indicated by the arrowhead) connects FM1 and FM2 (Figure 8.31b,c). Under this small magnetic field, the volume of the FM regions does not appreciably vary. Nevertheless, the

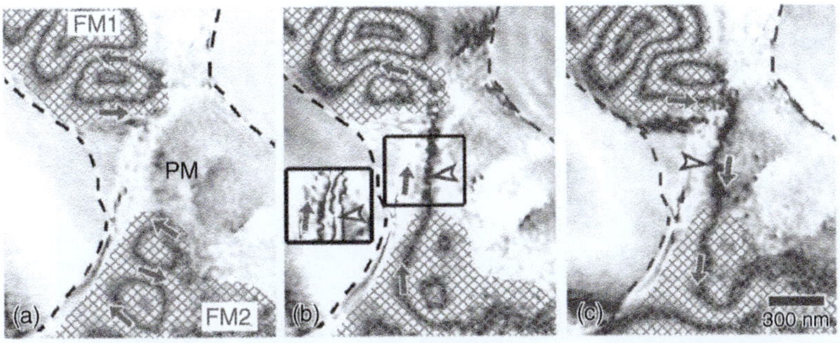

Figure 8.31 (a) Reconstructed phase image of separated ferromagnetic domains in $La_{0.81}Sr_{0.19}MnO_3$. Reconstructed phase images showing connection of ferromagnetic domains under an applied magnetic field of −41 mT (b) and with an applied magnetic field of 42 mT (c) (opposite direction of field applied in (b)). Inset in (b) is amplified phase information for rectangular area. Hatched portions represent area in which ferromagnetic regions are observed in absence of an applied magnetic field. Source: Yoo et al. [98], with permission from American Physical Society.

applied magnetic field appears to connect the separated FM domains by creating a magnetic flux channel. In addition, to elucidate the role of the observed flux channel in the CMR effect, researchers investigated the conductivity in the nanoscale channel using a double-probe piezodriving holder [82].

The effect of a thermal APB on ferromagnetic spin order is critical. We acquired Lorentz microscopy images for a wide temperature range, from 293 to 703 K of $Fe_{70}Al_{30}$ (Figure 8.32a–e). Heating the specimen to 573 K intensified the stripes (the meandering bright and dark lines), which originate from the deflection of incident electrons due to the Lorentz force [78, 95, 99, 100]. Importantly, these striations are present in the APB locations. The observations clearly indicate that a magnetic anomaly occurs in the APB region. Further heating reduced the magnetic contrast (Figure 8.32d,e) because of the elimination of spontaneous magnetization in this specimen. In fact, the magnetic contrast is almost invisible in Figure 8.32e, which only shows a faint (nonmagnetic) spotty pattern because of contaminants that appear to form at ∼700 K. The absence of magnetic signals at 703 K is consistent with magnetization measurements of bulk specimens [101, 102]; that is, the phase transition to the paramagnetic state was completed when the specimen was heated to ∼700 K.

Figure 8.32f–j presents two-dimensional mapping images of the phase gradient, observed as a function of temperature. The view field is identical to that shown in Figure 8.32a–e. The direction and magnitude of the phase gradient are represented on the basis of the color wheel shown in Figure 8.32j. As evident in Figure 8.32f, this area appears to be magnetized in approximately one direction indicated in green. The anomaly in the APB is not yet clear because of the substantial magnetization in both the matrix and APB regions at 293 K. Significant heating reduces the magnetization in the matrix, as demonstrated by the increase in size of the dark area in Figure 8.32g–i. Notably, magnetization in the APB remains pronounced (compared with the matrix region) at elevated temperatures (Figure 8.32h,i). These results

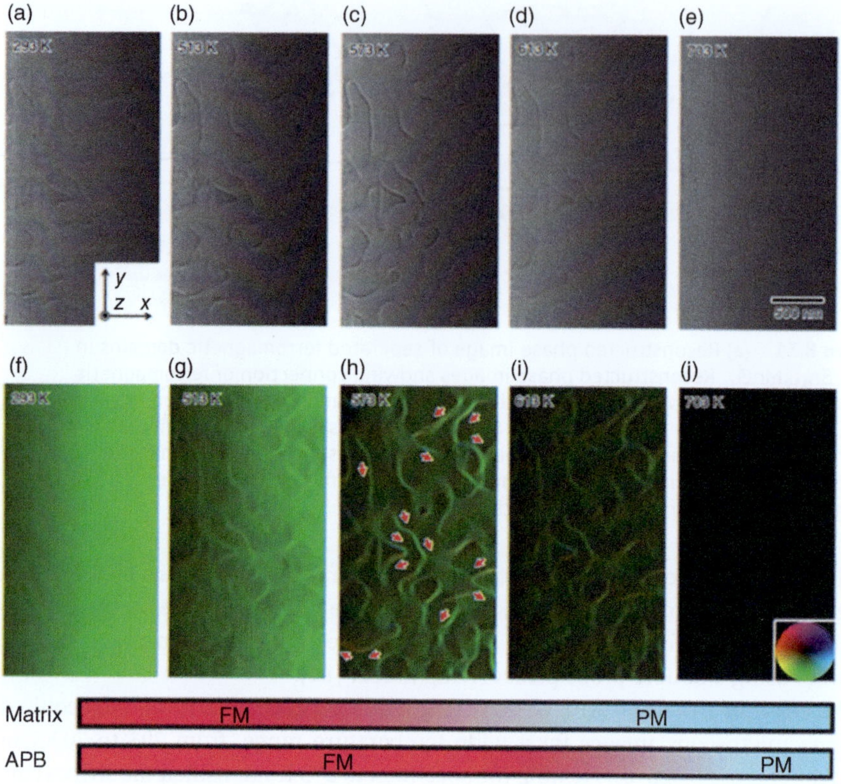

Figure 8.32 Direct imaging of magnetization distribution in APB region $Fe_{70}Al_{30}$. (a–e) Lorentz microscopy images acquired at 293, 513, 573, 613, and 703 K, respectively. Magnetic contrast (paired bright and dark lines) indicates that variation in local magnetic flux density can be observed along APBs: see images in (a–d). (f–j) Mapping of phase gradient revealed by electron holography. Results were collected from the same area as that shown in (a–e) at 293, 513, 573, 613, and 703 K, respectively. Phase gradient provides a measure of in-plane magnetic flux density, whose direction and magnitude are represented by color wheel shown in (j). An undesired contribution from mean inner potential was removed using a reference hologram acquired in paramagnetic phase at 713 K (not shown here). Undesired diffraction contrast was removed by tilting specimen when electron holograms were collected. FM and PM represent ferromagnetic and paramagnetic, respectively. Source: Murakami et al. (2014), with permission from Springer Nature.

definitively indicate that the ferromagnetic spin order is stabilized by the APBs. In addition, Figure 8.32h,i offers important information for understanding the ferromagnetic phase confined in the narrow, complex-shaped boundary region. In the case of low magnetocrystalline anisotropy in the cubic system, the magnetic dipolar interaction dominates the spin alignment in the APB. The magnetization vectors accordingly trace the highly meandering APB (refer to the positions labeled with red arrows indicating local magnetic flux directions in Figure 8.32h). As a result, the specimen shows a network of thin-layered ferromagnetic regions. Further heating renders the whole area of the specimen paramagnetic (Figure 8.32j). The effective

Curie temperature in the APB is likely higher than that of the matrix by at least 30 K. The locations of APBs remain unchanged upon heating to 703 K.

The spatial resolution of electron holography is noteworthy. The resolution can be approximated as three times the fringe pitch in an electron hologram, which is produced by interference between the reference wave and the object wave [100]. The fringe pitch was 17 nm for the results shown in Figure 8.32f–j; this value of the fringe pitch, which is larger than the APB width, was needed for image acquisition from a wide field of view (Figure 8.32). Because of this experimental condition, the observed APB regions are broadened in Figure 8.32f–i. Nevertheless, the observations clearly show the ferromagnetic phase that persists in the locations of APBs at elevated temperatures (see Figure 8.32h,i) because the electron holography study accurately determined the phase shift in the matrix regions, which are sufficiently larger (on the order of 100 nm) than the resolution. As demonstrated in Figure 8.33a–c, the magnetic anomaly in the APB region can be recognized by the deviation of the baselines (that is, plots of the phase shift) determined for the neighboring matrix regions.

Measurements of magnetic flux density were also carried out at elevated temperatures (573 and 653 K), as shown in Figure 8.33b,c. The values of the magnetic flux density measured for both the APB and matrix (Figure 8.33e) are plotted as a function of temperature in Figure 8.33d. The error bar (±0.08 T) is mainly due to the uncertainty associated with the specimen thickness measurement. Although the magnetic flux density in the matrix was reduced to 0.1 T or smaller at these elevated temperatures, the APB region continued to show substantial magnetic flux density: 0.78 T at 573 K and 0.58 T at 653 K. These results reasonably explain the magnetic flux maps, such as those imaged in Figure 8.32h,i, in which the ferromagnetic spin order persists in the narrow APB region. Because of time-consuming high-resolution electron holography experiments needed at elevated temperatures, we could determine the magnetic flux density of the APB only at three temperature points: 293, 573, and 653 K. In future experiments, it will be interesting to accurately determine the Curie temperature for both the APB and matrix regions on the basis of electron holography observations. These observations will also provide useful information for understanding the steps observed in the thermomagnetization curves in deformed Fe–Al alloys, including disordered regions [103–105]. Nevertheless, the results in Figures 8.32f–j and 8.33d explicitly indicate that the effective Curie temperature in the APB regions will be greater than that for the matrix.

In a nanocrystalline soft magnetic material of the form Fe–M–B (where M is Zr, Nb, or Hf), which is a so-called "Nanoperm," the magnetic flux distribution has been observed to change upon cooling (Figure 8.34) [106]. In the figure, two characteristic features are observed. One feature is the increase in the magnetic flux density; that is, the spacing of the contour lines becomes smaller with decreasing temperature. The other characteristic feature is the disappearance of magnetic ripples: a small fluctuation in the lines of magnetic flux disappears with decreasing temperature. The coercivity of the specimen is considered to result mainly from the difference in the saturation magnetization between the crystalline phase and the amorphous phase.

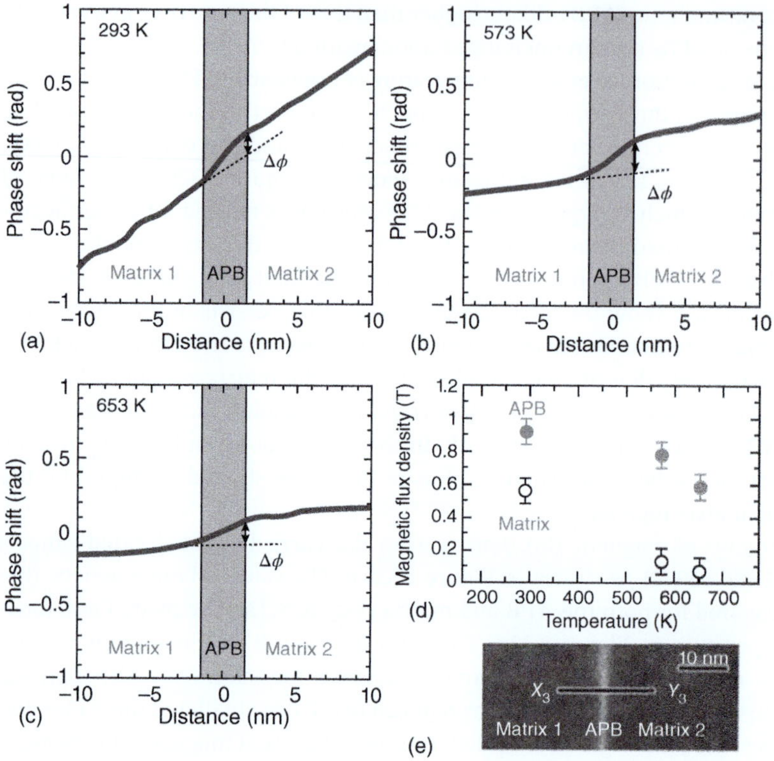

Figure 8.33 Determination of the magnetic flux density in APB region. (a–c) Plots of phase shift measured along X_3-Y_3 line (shown in e) at 293, 573, and 653 K, respectively. The results were acquired from the same specimen used in Figure 8.32. Blue and red curves show matrix and APB regions, respectively. $\Delta\varphi$ is phase difference (between observation and fitting curve for matrix 1 [dotted line]) observed at terminal of APB region. (d) Temperature dependence of magnetic flux density determined for APB (red) and matrix (blue). Error bars (±0.08 T) are mainly due to uncertainty in measuring specimen thickness. (e) TEM image (bright-field image) revealing location of APB. View field is identical to that indicated by rectangle in Figure 1b. Source: Murakami et al. (2014) with permission from Springer Nature.

The last topic of temperature-induced magnetic structure changes is high-temperature superconductors. In type-II superconductors, the critical current density J_c is the most important property related to their practical use because a high J_c can be achieved through pinning interaction between the quantized magnetic flux (fluxons) infiltrating the superconductor and the pinning centers [107, 108]. Improving the properties of superconductors requires understanding their magnetic flux behavior. Therefore, we combined electron holography and scanning ion microscopy (SIM) to visualize the magnetic flux distributions in the high-temperature superconductor Y–Ba–Cu–O with non-superconducting particles [109].

The bulk material was fabricated via a quench-and-melt growth process and consisted of a superconductive $YBa_2Cu_3O_{7-y}$ matrix (123 phase) and

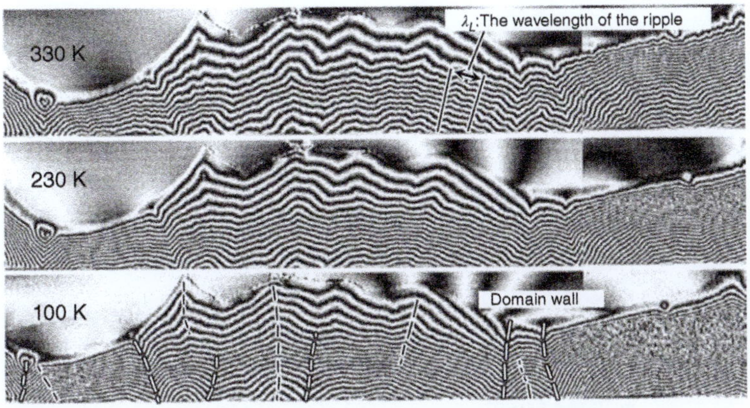

Figure 8.34 Reconstructed phase images of $Fe_{84}Nb_7B_9$ annealed at 773 K as a function of temperature; λ_L is longitudinal wavelength of magnetic ripple. Source: Gao et al. [106], with permission from AIP Publishing LLC.

non-superconductive Y_2BaCuO_5 particles (211 phase) [110, 111]. A small square-column specimen (2 μm × 2 μm × 25 μm) was cut from the bulk material using an FIB, with the *c*-axis perpendicular to the long axis of the square-column shape. The specimen contained one or two 211-phase particles in the direction of the incident electron beam.

The lower-left diagram in Figure 8.35 shows a schematic of the setting for specimen analysis. The specimen was fixed on the side edge of a semicircular Cu grid by tungsten deposition using an FIB. The *c*-axis of the 123 phase was aligned horizontally, and an external magnetic field was applied along the *c*-axis. Electron holography was performed using a 300 kV transmission electron microscope equipped with a liquid-He cryogenic system with magnetic coils for applying an external magnetic field [112]. The specimen holder was inserted above the pole pieces of the objective lens to reduce the magnetic field at the specimen position and thus preserve the magnetic structure in the specimen. During reconstruction of the phase images, the phase information above the critical temperature was subtracted to remove phase modulation because of the charging effects of the insulating layers on the surface of the specimen; these layers were introduced during the FIB thinning process.

The electron holography observations were performed in a magnetic field of 0.01 T. When the phase reconstruction process was performed under a constant magnetic flux applied to both the object wave and the reference wave, the reconstructed phase image showed a deviation of the magnetic flux from the zero field. To visualize the magnetic flux distribution around the particle, adding a constant magnetic flux in the specimen plane with a thickness corresponding to the particle diameter was useful. The phase of the reconstructed phase image was amplified by a factor of four. The center of Figure 8.35 shows a reconstructed phase image of a square-column-shaped Y–Ba–Cu–O specimen. The phase image was overlapped with the SIM image of the 211-phase particles to facilitate comparison. Because the heavier components in the SIM images are darker [113], the 211 phase exhibits darker contrast than the 123 phase, making identification of the 211-phase particles easier. We found that the

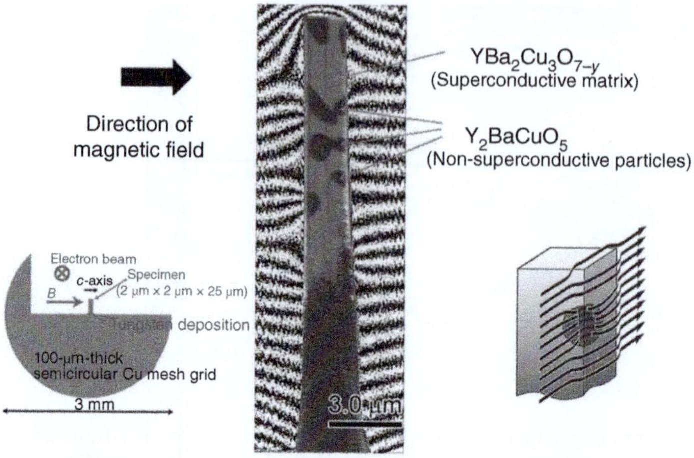

Figure 8.35 (left) Schematic of experimental setting. Specimen was fixed on side edge of a semicircular Cu mesh grid, and *c*-axis of superconductive phase was set parallel to external magnetic field. (center) Reconstructed phase image of a square-column-shaped Y–Ba–Cu–O specimen in an external magnetic field of 0.01 T at 12 K. Phase was amplified by a factor of 4. Phase image is overlain with a SIM image of specimen to enable comparison between magnetic flux distributions and 211-phase particles. (right) Schematic of magnetic flux lines around a specimen with a 211-phase particle located at specimen edge. Source: Shindo and Akase (2020), with permission from Elsevier.

lines of magnetic flux detoured around the top-edge region of the specimen because of the Meissner effect and that they were denser at the position of the 211-phase particles, as clarified from side-view SIM images of the specimen (Figure 8.36a,c) [109]. The lower-right diagram in Figure 8.35 is a schematic of the lines of magnetic flux around a specimen with a 211-phase particle located at the specimen edge.

The remanent state was examined by removing the external magnetic field, which led to the results shown in Figure 8.36. The reconstructed phase image (amplified by a factor of two) in Figure 8.36b shows the distribution of magnetic flux lines outside the specimen at 13 K; this image overlaps the SIM image of the front of the specimen. The reference wave was not affected by the external magnetic field because the field was turned off; therefore, the contour lines can be considered a projection of the magnetic flux lines. Because the phase was amplified by a factor of two, the magnitude of one fluxon ($h/2e$) corresponds to one period of the contour lines [109, 114, 115]. Figure 8.36a,c displays SIM images of the left- and right-side views, respectively, which show that the fluxons were stably pinned exactly at the 211-phase particle sites. This pinning occurred because these particles in the middle of the specimen were surrounded by the superconducting 123 phase. As demonstrated in this study, the combination of electron holography for characterizing the magnetic flux distribution and SIM imaging for characterizing the microstructure is useful for clarifying the pinning positions of magnetic flux. This technique can be extended to analyze the interactions between the magnetic flux and various lattice defects to clarify and improve the J_c of high-temperature superconductors.

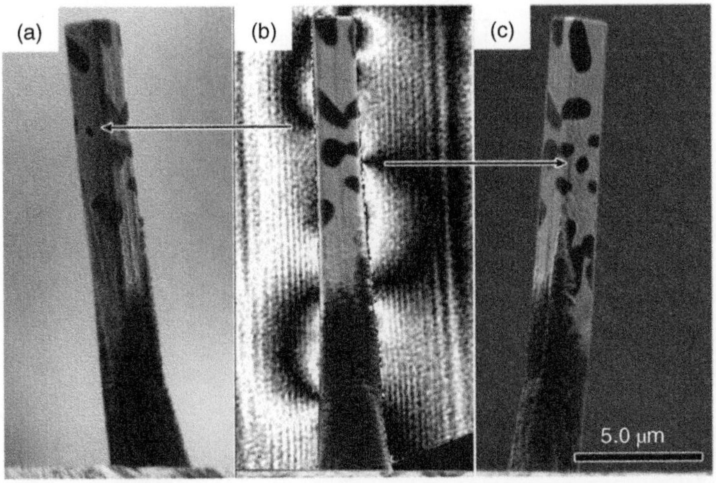

Figure 8.36 SIM and reconstructed phase images of a square-column-shaped Y–Ba–Cu–O superconductor in remanent state. SIM images in (a) and (c) are left- and right-side views of specimen. In (b), reconstructed phase image at 13 K is superimposed on SIM image to enable comparison between magnetic flux distributions and 211-phase particles. Emission points of fluxons in (b) correspond to insulating particles indicated by arrows in (a) and (c). Source: Akase et al. [109], with permission from AIP Publishing LLC.

8.4 Analysis of Three-Dimensional Magnetic Structures

In Section 8.2.3, the magnetic vortex structure of a skyrmion lattice was analyzed as a function of specimen thickness. In that experiment, a thin film with a step-like thickness change was prepared and analyzed from reconstructed phase images obtained with a high-voltage electron microscope.

The three-dimensional spin structure of the magnetic vortex of FeSiB, an amorphous soft magnetic material, has been investigated by holographic observation and computer simulation. A simple sample-tilting method was used, as shown in Figure 8.37, which includes a schematic of the magnetization vectors near the vortex center. The x–y plane indicates the disk plane with the y-axis as the tilting axis, and the z-axis indicates the direction of electron beams. Magnetization vectors are given as red arrows and are tilted upward in the z-direction. The magnetization rotation direction was counterclockwise (CCW) (chirality, $C = +1$), which was determined from the defocused Lorentz images under defocus conditions. At the center O, the magnetization vector stands up completely along the electron beam direction, indicating no in-plane magnetization component. We define this point as the phase-image center because the gradient of the phase shifts is zero. At point A, the magnetization vector is located in the x–z plane, with the tilting angle α measured from the z-axis. When we tilt the sample by α (in the CCW direction) while taking the y-axis as the axis of rotation, the direction of the magnetization vector at point A becomes parallel to the electron beam direction; thus, the phase-image center moves to A from O. At point B, the magnetization vector is tilted by $-\alpha$ from

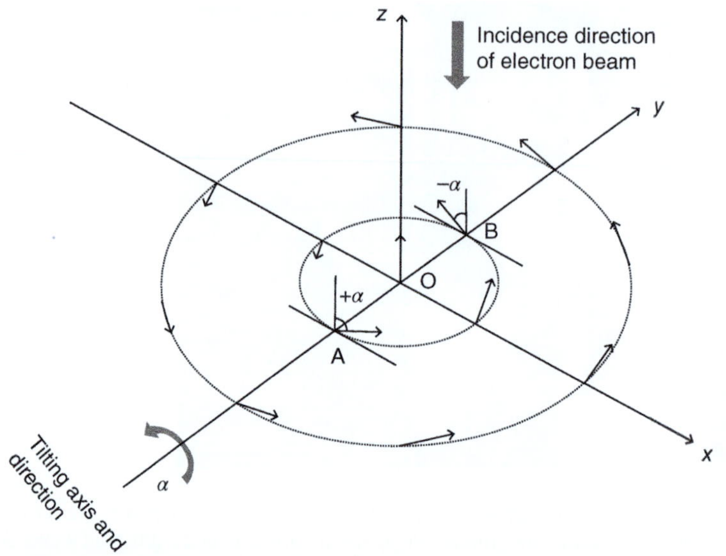

Figure 8.37 Schematic configuration of magnetization vectors near the vortex center O. The incident direction of the electron beam is taken as the $-z$-direction. The tilting angles of magnetization vectors measured from the z-axis at A and B are $+\alpha$ and $-\alpha$, respectively.

the z-axis. Thus, when we tilt the sample by $-\alpha$ (in the clockwise direction), the direction of the magnetization vector at point B becomes parallel to the electron beam direction, and the phase-image center moves to B. By tilting the specimen in this manner, we can estimate the distribution of magnetization on the basis of the shift of the phase-image center.

Using this idea, we carried out sample-tilting experiments. Three phase images obtained with a phase magnification of 1.0 are shown in Figure 8.38a–c with tilting angles −45°, 0°, and 45°, respectively. The three coordinates are indicated in the figure, where the rotation axis (y-axis) is parallel to the supporting handle of the specimen. Line-phase profiles are shown in Figure 8.38d along the red line for Figure 8.38a, black line for Figure 8.38b, and blue line for Figure 8.38c. The positions of the line profiles were determined by the following two rules: (i) the line should pass through the disk center and should be parallel to the support handle direction, and (ii) the line should be the centerline of the circular disk for 0° tilt and of the oval disk for 45° tilts. The line positions determined using either rule were found to be approximately the same. The profiles are aligned to make the zero-phase-shift points for the specimen edge located at the same position at ~500 nm, as indicated in Figure 8.38d. The enlarged figure of the rectangular part in Figure 8.38d is shown in Figure 8.38e, where the shifts of the phase-image center are clearly shown. The shift values are 10.1 ± 0.9 nm for the −45° tilt and 9.6 ± 0.8 nm for the 45° tilt, as determined by taking into account line location errors. This shift direction indicates

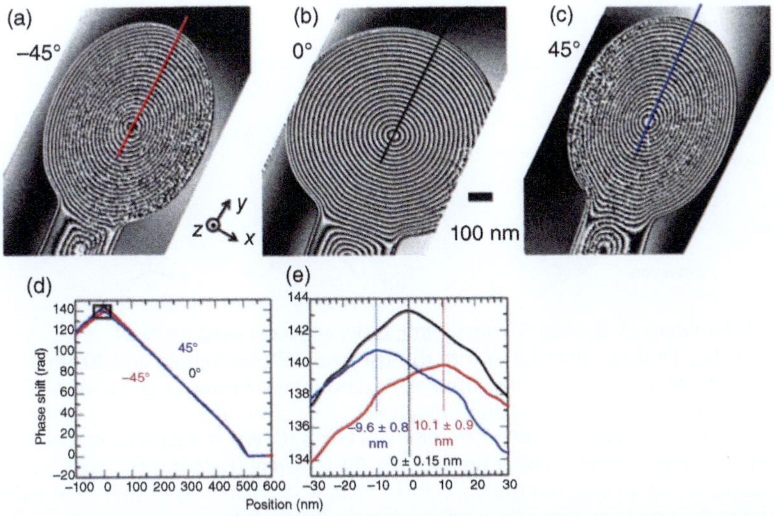

Figure 8.38 Reconstructed phase images of amorphous FeSiB with an amplification of 1.0 when specimen is tilted with respect to *y*-axis: (a) −45°, (b) 0°, and (c) 45°. (d) Profiles of phase shift along lines in (a–c). (e) Magnified profiles of phase shifts in rectangle in (d). In (d) and (e), position = 0 is set at phase-image center of (b). Source: Xia et al. (2012), with permission from Oxford University Press.

that the polarity of the specimen is +1 (magnetization in the +z-direction), as shown in Figure 8.37. The shifts of the phase-image center caused by the specimen tilting were simulated by considering both the internal magnetization and the external stray fields. The commercial software JMAG-Studio, which is based on the three-dimensional finite-element method, was used [116, 117]. A detailed comparison between the experimental data and simulation results was carried out [118].

To carry out three-dimensional magnetic structure analysis in detail, experiments with a simple thickness change or specimen tilting around a single tilting axis are insufficient. Rather, systematic three-dimensional magnetic structure analysis requires a specimen-tilting system with at least two rotation axes. The remaining component of the magnetic structure can then be calculated using one of Maxwell's equations:

$$\mathrm{Div}\vec{\mathrm{B}} = 0 \tag{8.1}$$

As reported elsewhere [119], three-dimensional magnetic structure analysis was performed by combining a high-voltage microscope and a dual-axis 360° rotation specimen holder. Figure 8.39a shows a schematic of the specimen structure of stacked ferromagnetic discs. Two Fe discs, 20 and 30 nm thick, were separated by a 10 nm Cr layer and sandwiched by a Cr substrate disc. The magnetization distribution depends on the balance between the demagnetization energy and the exchange energy. To analyze the three-dimensional magnetic structure using

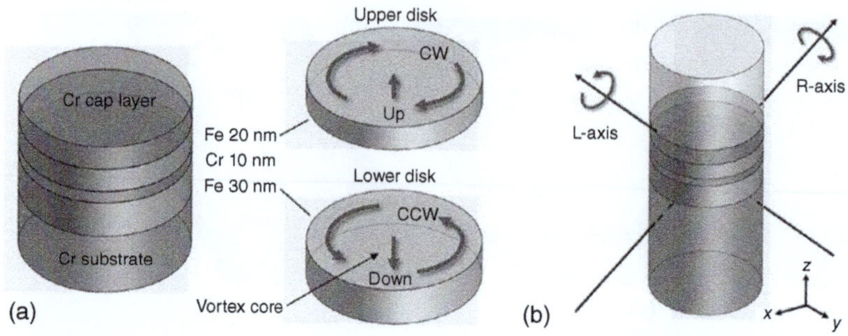

Figure 8.39 Schematics of stacked ferromagnetic disks and axes used for 360° observations. (a) Two Fe disks (light blue) with thicknesses of 20 nm (upper) and 30 nm (lower) were separated by a 10 nm Cr disk (green). (001)-oriented Fe and Cr disks were grown on a (001)-oriented Cr substrate and capped with a Cr layer. Direction of magnetic vortices in disk is either clockwise (CW) or counter clockwise (CCW). At magnetic vortex core, direction of magnetization is either up or down. (b) To reveal three-dimensional magnetic structure using an electron tomography technique, a pillar-shaped specimen with a carbon (gray) protection layer was observed by electron holography while specimen was rotated 360° around *R*- and *L*-axes. Two rotations were set perpendicular to each other in *x*–*z* plane. Electron beam was irradiated in *y*-direction.

a tomographic technique, a pillar-shaped specimen was observed with electron holography by rotating the specimen 360° around the *R*- and *L*-axes, as shown in Figure 8.39b. The two rotations were set perpendicular to each other in the *x*–*z* plane. The direction of the incident electron beam was then parallel to the *y*-direction.

Figure 8.40 shows the magnetic phase shifts obtained at specimen rotation angles of 0°, 60°, and 80° around the *R*-axis. The lower figures schematically show the specimen at each observation angle; the upper figures show the phase shifts in the form of $\cos(n\phi_M)$, where n is the phase amplification factor, which was set to 15. The black contour lines correspond to the projected components of the magnetic flux perpendicular to the electron beam. Contour lines were not observed outside the sample, indicating that the magnitude of magnetic flux leaking from the sample was small and did not affect the analyses. At 0°, the projected specimen thickness was a maximum of 250 nm. The obtained magnetic phase shifts, however, clearly show a projected magnetic flux (Figure 8.40a). These results demonstrate the advantages of using a 1 MV holography electron microscope for three-dimensional magnetic analysis.

Figure 8.41a shows a cross section of one of the three-dimensional magnetic fields. The direction of *z*-component of magnetic field at the vortex core of the upper disk was up, whereas that of the lower disk was down. These opposite *z*-directions are indicated by coloring in blue and red, respectively. Figure 8.41b shows a three-dimensional view of the magnetic structure at the vortex cores of the upper and lower discs. The tail-to-tail magnetic vortex cores were confirmed, and they were mutually repulsive. The obtained results were successfully compared with those of micromagnetic simulation based on the Landau–Lifshitz–Gilbert (LLG) equation.

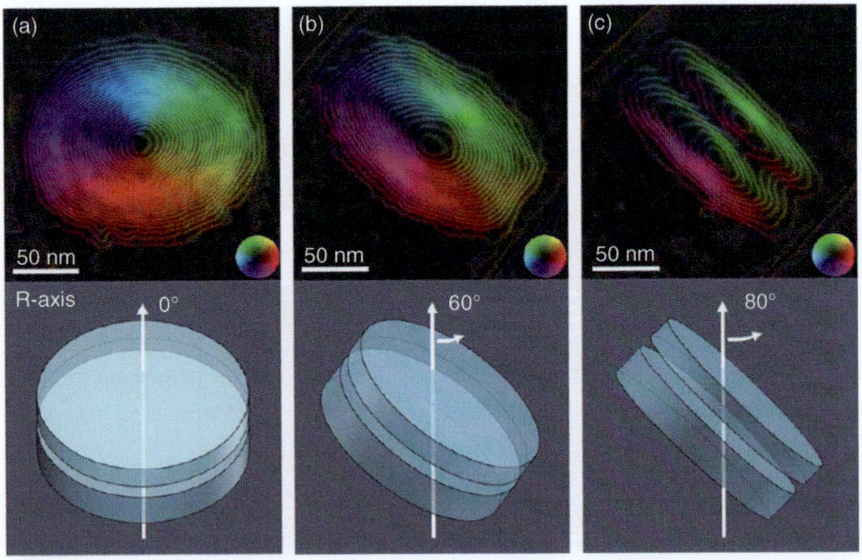

Figure 8.40 Typical magnetic phase shifts φ_M obtained by rotating specimen around R-axis, for rotation angles of (a) 0°, (b) 60°, and (c) 80°. Upper figures show $\cos(15\varphi_M)$ (magnetic phase shifts amplified 15 times) overlapped with color maps of projected magnetic vectors. Color wheels indicate directions of projected magnetic flux. Lower figures show schematics of specimen orientation with respect to electron beam direction, which was perpendicular to image plane.

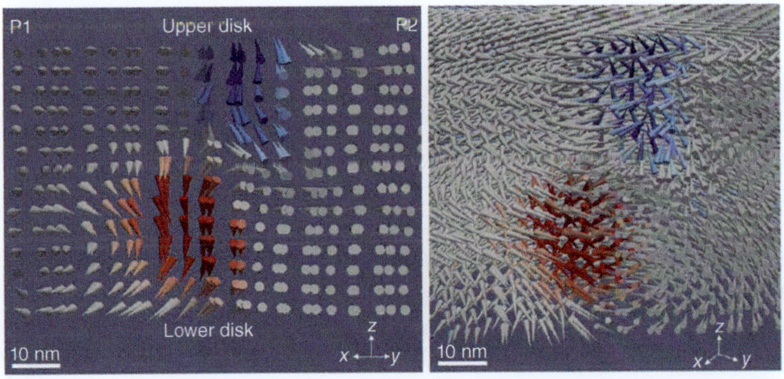

(a) (b)

Figure 8.41 Three-dimensional view of reconstructed magnetic vortex cores. z-direction components are indicated by blue (+z) or red (−z). (a) Cross-sectional magnetization distribution, which had opposite z-directions for upper and lower vortex cores. (b) Three-dimensional view of tail-to-tail vortex cores.

References

1 Tripp, S.L., Dunin-Borkowski, R.E., and Wei, A. (2003). *Angew. Chem. Int. Ed.* 115: 5749.

2 Sugawara, A., Fukunaga, K.-I., Scheinfein, M.R. et al. (2007). *Appl. Phys. Lett.* 91: 262513.

3 Hÿtch, M.J., Dunin-Borkowski, R.E., Scheinfein, M.R. et al. (2003). *Phys. Rev. Lett.* 91: 257207.

4 Gao, Y., Bao, Y., Pakhomov, A.B. et al. (2006). *Phys. Rev. Lett.* 96: 137205.

5 Gao, Y., Shindo, D., Bao, Y., and Krishnan, K. (2007). *Appl. Phys. Lett.* 90: 233105.

6 Biskupek, J., Kaiser, U., Lichte, H. et al. (2005). *J. Magn. Magn. Mater.* 293: 924.

7 Takeno, Y., Murakami, Y., Sato, T. et al. (2014). *Appl. Phys. Lett.* 105: 183102.

8 Tonomura, A., Matsuda, T., Endo, J. et al. (1986). *Phys. Rev. B: Condens. Matter* 34: 3397.

9 Kasama, T., Harrison, R.J., Church, N.S. et al. (2013). *Phase Transitions* 86: 67.

10 Chang, M.-T., Chou, L.-J., Hsieh, C.-H. et al. (2007). *Adv. Mater.* 19: 2290.

11 ELF Co. www.elf.co.jp.

12 Tonomura, A., Matsuda, T., Tanabe, H. et al. (1982). *Phys. Rev. B: Condens. Matter* 25: 6799.

13 Yoshizawa, Y., Oguma, S., and Yamauchi, K. (1988). *J. Appl. Phys.* 64: 6044.

14 Shindo, D., Park, Y.-G., and Yoshizawa, Y. (2002). *J. Magn. Magn. Mater.* 238: 101.

15 Visser, E.G. (1984). *J. Magn. Magn. Mater.* 42: 286.

16 Kondo, K., Chiba, T., Yamada, S., and Otsuki, E. (2000). *J. Appl. Phys.* 87: 6229.

17 Kasahara, T., Shindo, D., Yoshikawa, H. et al. (2007). *J. Electron Microsc.* 56: 7.

18 Hefferman, S.J., Chapman, J.N., and McVitie, S. (1991). *J. Magn. Magn. Mater.* 95: 76.

19 Uhlig, T., Heumann, M., and Zweck, J. (2003). *Ultramicroscopy* 94: 193.

20 Yi, G., Nicholson, W.A.P., Lim, C.K. et al. (2004). *Ultramicroscopy* 99: 65.

21 Cumings, J., Olsson, E., Petford-Long, A.K., and Zhu, Y. (2008). *MRS Bull.* 33: 101.

22 Akase, Z. and Shindo, D. (2010). *J. Electron Microsc.* 59: 207.

23 Arita, M., Tokuda, R., Hamada, K., and Takahashi, Y. (2014). *Mater. Trans.* 55: 403.

24 Grüendmayer, J. and Zweck, J. (2006). Proceedings of the sixteenth International Microscopy Congress. Sapporo, Japan (3–8 September 2006), 1541.

25 Akase, Z., Shindo, D., Inoue, M., and Taniyama, A. (2007). *Mater. Trans.* 48: 2626.

26 Akase, Z., Aizawa, S., Shindo, D. et al. (2015). *J. Magn. Magn. Mater.* 375: 10.

27 Croat, J.J., Herbst, J.F., Lee, R.W., and Pinkerton, F.E. (1984). *J. Appl. Phys.* 55: 2078.

28 Sagawa, M., Fujimura, S., Togawa, N. et al. (1984). *J. Appl. Phys.* 55: 2083.

29 McCartney, M.R. and Zhu, Y. (1998). *J. Appl. Phys.* 83: 6414.

30 Zhu, Y. and McCartney, M.R. (1998). *J. Appl. Phys.* 84: 3267.

31 Park, Y.-G. and Shindo, D. (2002). *J. Magn. Magn. Mater.* 238: 68.

32 Cho, Y., Sasaki, T., Harada, K. et al. (2020). *Scr. Mater.* 178: 533.

33 Kojima, H. and Wohlfarth, E.P. (ed.) (1982). *Ferromagn. Mater.* 3: 305.

34 Aiso, T., Shindo, D., and Sato, T. (2007). *J. Magn. Magn. Mater.* 318: 18.

35 Okabe, F., Park, H.S., Shindo, D. et al. (2006). *Mater. Trans.* 47: 218.
36 Shindo, D., Park, Y.-G., Gao, Y., and Park, H.S. (2004). *J. Appl. Phys.* 95: 6521.
37 Kim, J.J., Park, H.S., Shindo, D. et al. (2006). *Mater. Trans.* 47: 907.
38 Cullity, B.D. (1972). *Introduction to Magnetic Materials*. Addison-Wesley.
39 Wohlfarth, E.P. (1982). *Ferromagnetic Materials*, vol. 3, 107. North-Holland Publishing Company.
40 O'Handley, R.C. (1999). *Modern Magnetic Materials*, 476. Wiley.
41 Shindo, D. and Oikawa, T. (2002). *Analytical Electron Microscopy for Materials Science*, 111. Springer.
42 Hütten, A., Reiss, G., Saikaly, W., and Thomas, G. (2001). *Acta Mater.* 49: 827.
43 Hubert, A. and Schafer, R. (1998). *Magnetic Domains*, vol. 550. Springer.
44 Park, H.S., Park, Y.-G., Gao, Y. et al. (2005). *J. Appl. Phys.* 97: 033908.
45 Dunin-Borkowski, R.E., McCartney, M.R., Kardynal, B., and Smith, D.J. (1998). *J. Appl. Phys.* 84: 374.
46 Heumann, M., Uhlig, T., and Zweck, J. (2005). *Phys. Rev. Lett.* 94: 077202.
47 Park, J.-B., Park, G.-S., Song, I.-Y. et al. (2006). *Microscopy* 55: 17.
48 Kasama, T., Barpanda, P., Dunin-Borkowski, R.E. et al. (2005). *J. Appl. Phys.* 98: 013903.
49 Furrer, S., Lantz, M.A., Reininger, P. et al. (2018). *IEEE Trans. Magn.* 54: 3100308.
50 Osakabe, N., Yoshida, K., Horiuchi, Y. et al. (1983). *Appl. Phys. Lett.* 42: 746.
51 Tonomura, A. (1999). *Electron Holography*, 2e. Springer-Verlag.
52 Xia, W.X., Tohara, K., Murakami, Y. et al. (2006). *IEEE Trans. Magn.* 42: 3252.
53 Kim, J.J., Hirata, K., Ishida, Y. et al. (2008). *Appl. Phys. Lett.* 92: 162501.
54 Fert, A. and Grünberg, P. (2007). Prix Nobel Noble Prizes. (ed. K. Gd), The 2007 Nobel Prize in Physics: Albert Fert and Peter Grünberg.
55 Chappert, A., Fert, A., and Van Dau, F.N. (2007). *Nat. Mater.* 6: 813.
56 Dieny, B., Speriosu, V.S., Metin, S. et al. (1991). *J. Appl. Phys.* 69: 4774.
57 Moodera, J.S., Kinder, L.R., Wong, T.M., and Meservey, R. (1995). *Phys. Rev. Lett.* 74: 3273.
58 Goto, T., Jeong, J.S., Xia, W. et al. (2013). *Microscopy* 62: 383.
59 Park, H.S., Hirata, K., Yanagisawa, K. et al. (2012). *Small* 8: 3640.
60 Iwasaki, J., Mochizuki, M., and Nagaosa, N. (2013). *Nat. Commun.* 4: 1463.
61 Lee, M., Kang, W., Onose, Y. et al. (2009). *Phys. Rev. Lett.* 102: 186601.
62 Tonomura, A., Yu, X., Yanagisawa, K. et al. (2012). *Nano Lett.* 12: 1673.
63 Seki, S., Yu, X.Z., Ishiwata, S., and Tokura, Y. (2012). *Science* 336: 198.
64 Park, H.S., Yu, X., Aizawa, S. et al. (2014). *Nat. Nanotechnol.* 9: 337.
65 Mathur, N., Stolt, M.J., Niitsu, K. et al. (2019). *ACS Nano* 13: 7833.
66 Shibata, K., Tanigaki, T., Akashi, T. et al. (2018). *Nano Lett.* 18: 929.
67 Ullakko, K., Huang, J.K., Kantner, C. et al. (1966). *Appl. Phys. Lett.* 1996: 69.
68 Müllner, P., Chernenko, V.A., and Kostorz, G. (2003). *Scr. Mater.* 49: 129.
69 James, R.D. and Wuttig, M. (1998). *Philos. Mag.* 77: 1273.
70 Kakeshita, T., Takeuchi, T., Fukuda, T. et al. (2000). *Appl. Phys. Lett.* 77: 1502.
71 Oikawa, K., Wulff, L., Iijima, T. et al. (2001). *Appl. Phys. Lett.* 79: 3290.
72 Murakami, Y., Shindo, D., Oikawa, K. et al. (2002). *Acta Mater.* 50: 2173.

73 Oikawa, K., Ota, T., Ohmori, T. et al. (2002). *Appl. Phys. Lett.* 81: 2360.

74 De Graef, M., Willard, M.A., McHenry, M.E., and Zhu, Y. (2001). *IEEE Trans. Magn.* 37: 2663.

75 Tsuchiya, K., Yamamoto, K., Hirayama, T., et al. (2003) Proceedings of the TMS 2003 Annual Meeting. San Diego, CA. 2003 TMS Annual Meeting & Exhibition: Michael Meshii Honarary Dinner, 305.

76 Pan, Q. and James, R.D. (2000). *J. Appl. Phys.* 87: 4702.

77 Chopra, H.D., Ji, C., and Kokorin, V.V. (2000). *Phys. Rev. B: Condens. Matter* 61: R14913.

78 Venkateswaran, S.P., Nuhfer, N.T., and De Graef, M. (2007). *Acta Mater.* 55: 2621.

79 Venkateswaran, S.P., Nuhfer, N.T., and De Graef, M. (2007). *Acta Mater.* 55: 5419.

80 Murakami, Y., Shindo, D., Oikawa, K. et al. (2003). *Appl. Phys. Lett.* 82: 3695.

81 Murakami, Y., Shindo, D., Kobayashi, K. et al. (2006). *Mater. Sci. Eng., A* 438–440: 1050.

82 Murakami, Y., Yano, T., Shindo, D. et al. (2007). *Metall. Mater. Trans. A* 38: 815.

83 Yano, T., Murakami, Y., Kainuma, R., and Shindo, D. (2007). *Mater. Trans.* 48: 2636.

84 Jin, S., Tiefel, T.H., McCormack, M. et al. (1994). *Science* 264: 413.

85 Tokura, Y., Urushibara, A., Moritomo, Y. et al. (1994). *J. Phys. Soc. Jpn.* 63: 3931.

86 Moreo, A., Yunoki, S., and Dagotto, E. (1999). *Science* 283: 2034.

87 Fäth, M., Freisem, S., Menovsky, A.A. et al. (1999). *Science* 285: 1540.

88 Uehara, M., Mori, S., Chen, C.H., and Cheong, S.-W. (1999). *Nature* 399: 560.

89 Tokunaga, M., Tokunaga, Y., and Tamegai, T. (2004). *Phys. Rev. Lett.* 93: 037203.

90 Zhang, L., Israel, C., Biswas, A. et al. (2002). *Science* 298: 805.

91 Lecoeur, P., Trouilloud, P.L., Xiao, G. et al. (1997). *J. Appl. Phys.* 82: 3934.

92 Fukumura, T., Sugawara, H., Hasegawa, T. et al. (1969). *Science* 1999: 284.

93 Mori, S., Asaka, T., and Matsui, Y. (2002). *J. Electron Microsc.* 51: 225.

94 Loudon, J.C., Mathur, N.D., and Midgley, P.A. (2002). *Nature* 420: 797.

95 Murakami, Y., Yoo, J.H., Shindo, D. et al. (2003). *Nature* 423: 965.

96 Yoo, J.H., Murakami, Y., Shindo, D., and Ohsuna, T. (2004). *Philos. Mag.* 84: 2597.

97 Yoo, J.H., Murakami, Y., Shindo, D. et al. (2002). *Phys. Rev. B: Condens. Matter* 66: 212406.

98 Yoo, J.H., Murakami, Y., Shindo, D. et al. (2004). *Phys. Rev. Lett.* 93: 047204.

99 Murakami, Y., Yanagisawa, K., Niitsu, K. et al. (2013). *Acta Mater.* 61: 2095.

100 Völkl, E., Allard, L.F., and Joy, D.C. (1999). *Introduction to Electron Holography*. New York: Kluwer Academic/Plenum Publishers.

101 Rodríguez, D.M., Plazaola, F., Garitaonandia, J.S. et al. (2012). *Intermetallics* 24: 38.

102 Noakes, D.R., Arrott, A.S., Belk, M.G. et al. (2004). *J. Appl. Phys.* 95: 6574.

103 Hernando, A., Amils, X., Nogués, J. et al. (1998). *Phys. Rev. B: Condens. Matter* 58: R11864.

104 Amils, X., Nogués, J., Suriñach, S. et al. (2001). *Phys. Rev. B: Condens. Matter* 63: 052402.

105 Apinaniz, E., Garitaonandia, J.S., Plazaola, F. et al. (2003). *J. Magn. Magn. Mater.* 254, 255: 136.

106 Gao, Y., Shindo, D., Bitoh, T., and Makino, A. (2003). *J. Appl. Phys.* 93: 7462.

107 Campbell, A.M. and Evetts, J.E. (1972). *Adv. Phys.* 21: 199.

108 Plain, J., Puig, T., Sandiumenge, F. et al. (2002). *Phys. Rev. B: Condens. Matter* 65: 104526.

109 Akase, Z., Kasai, H., Mamishin, S. et al. (2012). *J. Appl. Phys.* 111: 033912.

110 Morita, M., Sawamura, M., Takebayashi, S. et al. (1994). *Physica C* 235–240: 209.

111 Morita, M., Kumada, M., Sato, A. et al. (2007). *Physica C* 463–465: 387.

112 Harada, K., Endo, J., Osakabe, N., and Tonomura, A. (2008). *e-J. Surf. Sci. Nanotech.* 6: 29.

113 Kim, K.H., Akase, Z., Suzuki, T., and Shindo, D. (2010). *Mater. Trans.* 51: 1080.

114 Lichte, H. (2008). *Ultramicroscopy* 108: 256.

115 Shindo, D. and Murakami, Y. (2008). *J. Phys. D: Appl. Phys.* 41: 183002.

116 Shibata, N., Kohno, Y., Nakamura, A. et al. (2019). *Nat. Commun.* 10: 2308.

117 Joy, D.C., Roming, A.D., and Goldstein, J. (1986). *Principles of Analytical Electron Microscopy*. Springer Science & Business Media.

118 Xia, W., Aizawa, S., Tanigaki, T. et al. (2012). *J. Electron Microsc.* 61: 71.

119 Tanigaki, T., Takahashi, Y., Shimakura, T. et al. (2015). *Nano Lett.* 15: 1309.

Part IV

Visualization of Collective Motions of Electrons and Their Interpretation

So far, we have described various in situ experiments to observe electric and magnetic fields by electron holography. In the analyses of electric fields of toner particles and organic photoconductors presented in Sections 6.6 and 7.4, respectively, the charging effects due to the high-energy electron irradiation were suppressed by metallic shields set in the piezodriving probes. The magnitudes of the charging effects of insulators depend on the incident electron beam intensity, secondary electron intensity from substrates, and other factors. An effective way to control the charging effect is to coat insulating specimens with metallic elements.

Figure 1 shows the difference in charging effects depending on the amount of Os coating and carbon sputtering for a typical biological specimen, a microfibril of sciatic nerve tissue (mouse). Figure 1a shows a TEM image of sciatic nerve tissue, while Figure 1b shows one in which there is severe specimen drift. The difference results from the amount of Os coating and carbon sputtering, as indicated in the EDS spectra, which show that the relative intensities of characteristic X-rays from C and Os are much weaker than that from a supporting Cu plate, as shown in Figure 1b.

Controlling the charging effect due to the secondary electron emission, a new application of electron holography, i.e. the visualization of electrons, can be performed as explained in detail in the first half of Part IV. Visualizing the motions of electrons is particularly interesting because electromagnetic fields basically originate from the various motions of electrons. Studies using this application started with the charging effect on biological specimens. Subsequent studies investigated the accumulation and distribution of electrons for various insulating materials, with special attention paid to their surfaces.

Finally, we discuss the theoretical interpretation of the experimental results of visualization of electron motion at nanometer scale. We also present an interpretation of wave–particle dualism and a method for modeling electron interference effects.

Material Characterization using Electron Holography, First Edition. Daisuke Shindo and Takeshi Tomita.

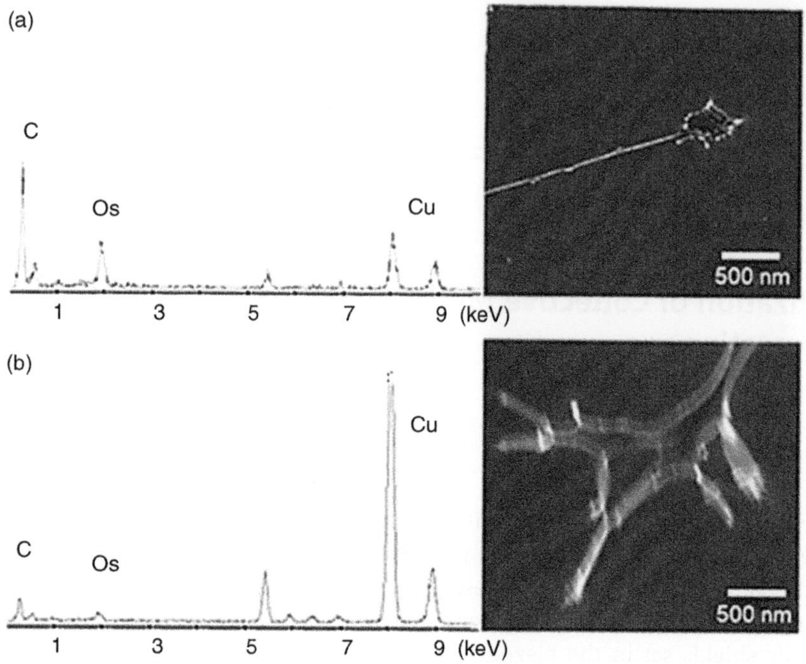

Figure 1 (a) TEM image of microfibril of sciatic nerve tissue with energy dispersive X-spectrum. (b) Microfibril with poor C and Os showing drift due to charging effect.

9

Charging Effects and Secondary Electron Distribution of Biological Specimens

Figure 9.1a shows schematic illustrations of the charging effect. When an insulating specimen such as a biological one is irradiated with high-energy electrons, secondary electrons are emitted, and the specimen becomes positively charged. When the electron irradiation intensity increases, as shown in (b), the specimen becomes strongly charged. Eventually, secondary electrons emitted from the specimen tend to be attracted to it. For some geometric configurations of the specimen surface and electric field, the secondary electrons tend to accumulate and/or form stationary motions.

The biological specimens used for the electron holography experiments described below were microfibrils of sciatic nerve tissue, obtained from adult C57BL/6 mice prepared by freeze-drying. The dried specimens were treated in OsO_4 vapor for two hours and further sputtered with carbon for 20 seconds [1]. The microfibrils were fixed with carbon tape to the copper plate of the specimen holder.

9.1 Visualization of Stationary Electron Orbits

9.1.1 Stationary Electron Orbits Observed Around Microfibrils

Figure 9.2 shows TEM images of a microfibril of sciatic nerve tissue observed under over-focus condition: the electron beam was focused under the specimen position. For a line-shaped insulating specimen, the incident electron beam was deflected by the positively charged specimen. As the intensity of the beam was increased, the contrast in the top-edge region changed drastically. As indicated in the inserted illustrations, this contrast change is interpreted as enhanced deflection of the incident electron beam due to the increased charging effect, especially in the top-edge region [2]. The stationary orbits of secondary electrons, as described below, result from this enhancement of the charging effects in the top-edge region in line-shaped specimens.

Figure 9.3a,c,d, respectively, shows a hologram, a reconstructed phase image, and a reconstructed amplitude image of three entangled microfibrils [3]. The intensity of the incident electron beam for observation of the hologram was $60\,\mathrm{nm^{-2}\,s^{-1}}$.

Material Characterization using Electron Holography, First Edition. Daisuke Shindo and Takeshi Tomita.

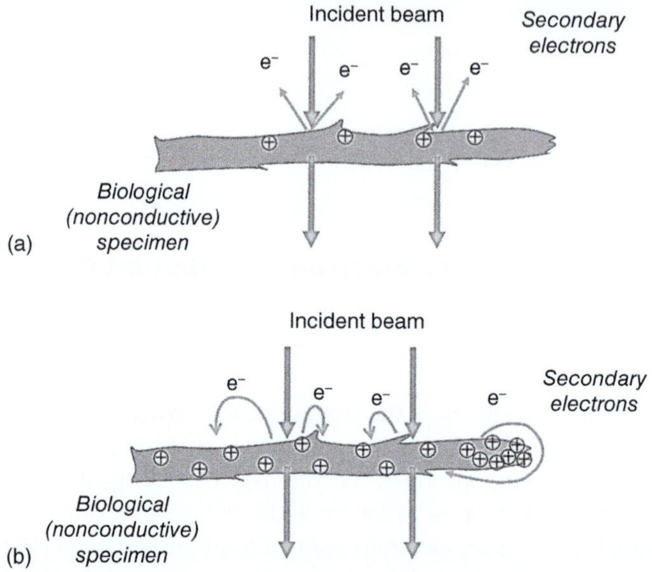

Figure 9.1 Schematic illustrations showing effect of charging on biological specimen: (a) initial state and (b) state after severe charging under strong electron beam irradiation.

The electric potential at the specimen surface was evaluated as 30 V through computer simulation [1]. In the reconstructed phase image (Figure 9.3c), there are several irregular contrasts resulting from electric field variation due to the motion of the secondary electrons. As explained in regard to Figure 9.8, the hologram has interference fringes missing in regions A, B, and D (Figure 9.3b), resulting in the background in the digital diffractograms having a large amplitude, whereas it has no missing fringes in region C.

The reconstructed amplitude image in Figure 9.3d shows several circuits with dark contrasts around the top-edge regions of the microfibrils, corresponding to the orbits of secondary electrons around these regions. Furthermore, in the enlarged hologram of region E, the extrapolated dotted lines along the interference fringes shift away from each other in the regions with no interference fringes. These characteristic features were clarified well by the simulation described below [3]. It was reported that the interference fringes in the hologram disappeared completely when the projected density of fluctuating secondary electrons was approximately 1 nm^{-2} for incident 200- and 300-keV electrons [3, 4].

Observation with a single microfibril was carried out for comparison with the stationary electron orbits around the tangled microfibrils. As shown by the bright-field image in Figure 9.4a, the surface of the microfibril was uneven, and there were some dark spots, which are indicated by blue arrowheads. Since the dark contrast is attributed to the existence of heavy elements, the dark spots are considered to be Os-rich regions.

The uneven surface and inhomogeneous distribution of the coating element would likely cause an inhomogeneous charge distribution on the surface of the

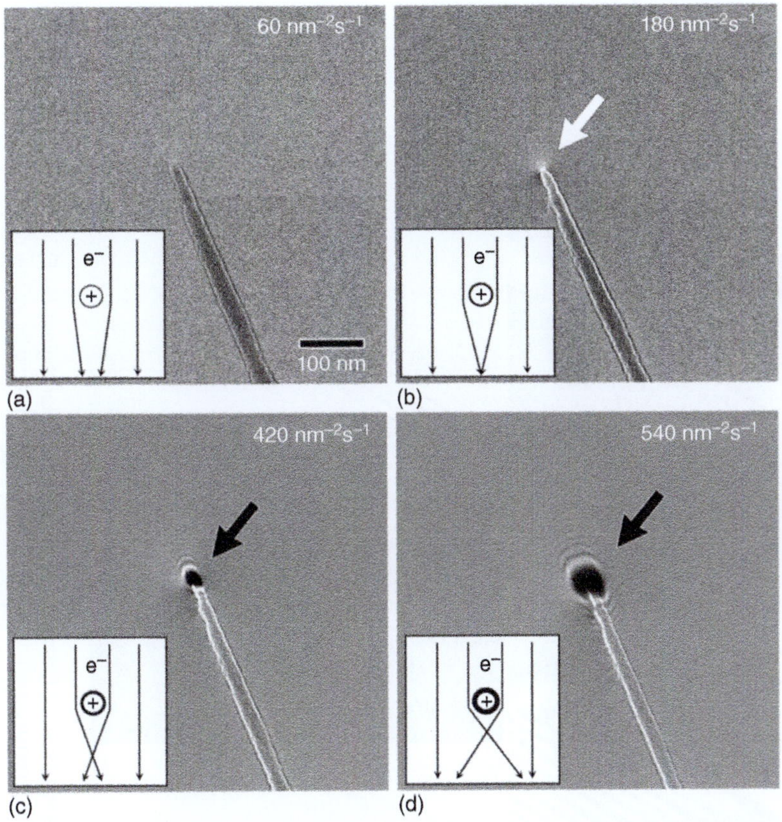

Figure 9.2 TEM images of microfibril of sciatic nerve tissue observed under over-focus condition. As intensity of incident electron beam was increased, contrast in top-edge region changed drastically. Inserted illustrations show deflection of incident electron beam due to charging effect in top-edge region. Source: Shindo and Akase(2020), with permission from ELSEVIER.

microfibrils of sciatic nerve tissue, resulting in the characteristic motions of collective secondary electrons. The reconstructed phase image in Figure 9.4b, which was obtained at an electron intensity of 30 nm^{-2} s^{-1}, shows an electric field around the microfibril caused by the charging effect. In the reconstructed amplitude image in (c), there is a slightly dark contrast on the right side of the microfibril, as indicated by a short yellow arrow. When the incident electron intensity was increased to 36 nm^{-2} s^{-1}, the dark-contrast region shifted to the left of the microfibril, as shown in (d).

It is interesting to note that a sharp dark-contrast region was observed along the microfibril indicated by the narrow red arrow; it is similar to that for the tangled microfibrils described below. Soon thereafter (within a few tens of seconds), as shown in Figure 9.4e, the dark-contrast region formed a circular shape around the top of the microfibril, which tends to be charged more strongly than other parts. The sharp line contrast was still visible, as indicated by the narrow red arrow.

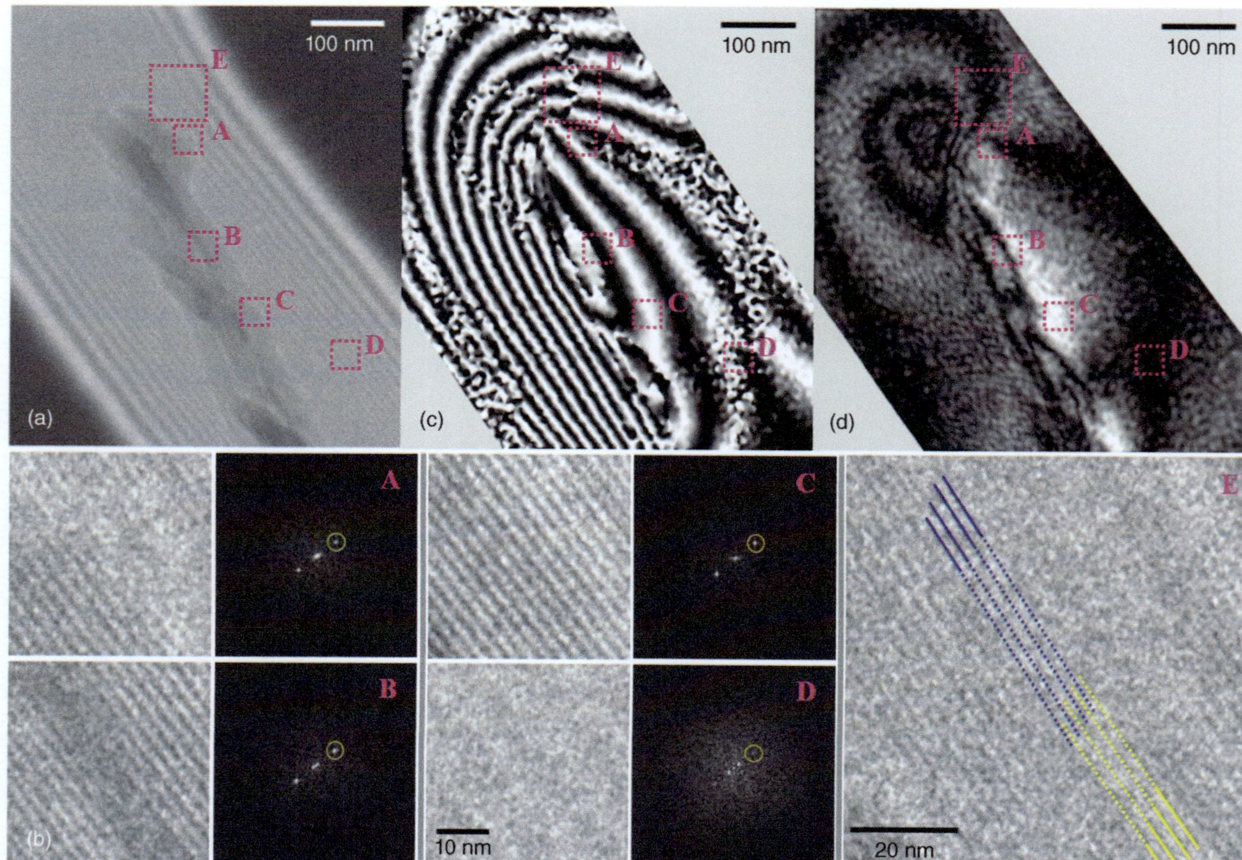

Figure 9.3 (a) Electron hologram of tangled microfibrils of sciatic nerve tissue. (b) Enlarged images of rectangular regions (A–E) indicated by broken pink lines in (a) together with digital diffractograms (A–D). Aperture size for reconstruction of amplitude image is indicated by yellow circle in each digital diffractogram. (c) Reconstructed phase image. (d) Reconstructed amplitude image. Rectangular regions (A–E) are also presented in (c) and (d) for comparison. Source: Shindo et al. (2009), with permission from The Physical Society of Japan.

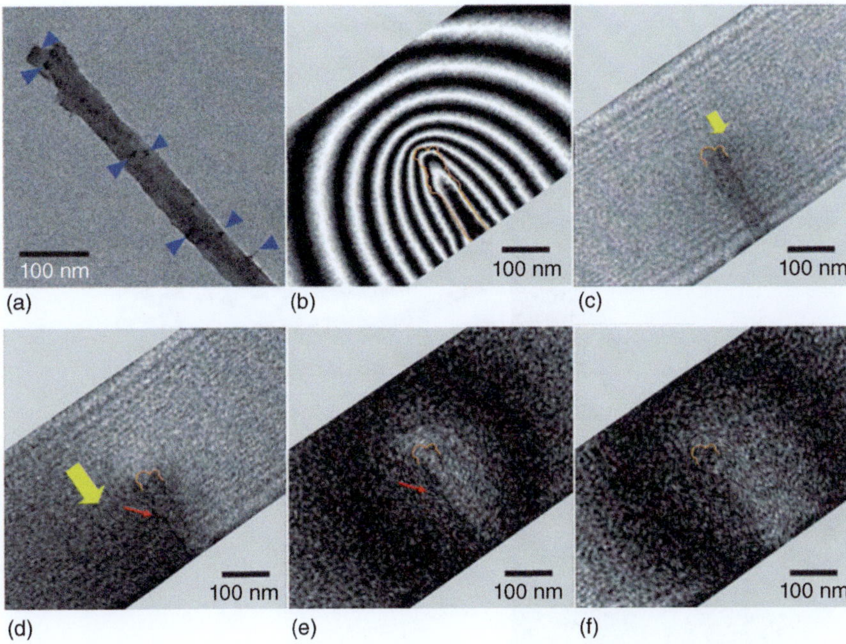

Figure 9.4 (a) Bright-field image of single microfibril of sciatic nerve tissue. (b and c) Reconstructed phase and amplitude images, respectively, of microfibril at incident electron intensity of 30 nm^{-2} s^{-1}. Corresponding parts of microfibril are outlined in brown. (d–f) Reconstructed amplitude images at incident electron intensity of 36 nm^{-2} s^{-1}. Source: Shindo et al. (2009), with permission from The Physical Society of Japan.

After another few tens of seconds, as shown in (f), the circular shape became larger and shifted to the right. Because the dark-contrast region is attributed to the orbits of secondary electrons, the change in its shape around a single microfibril indicates that the rather unstable motion of collective secondary electrons differs from those observed with tangled microfibrils.

When the incident electron intensity was increased further after observing the holograms of microfibrils showing the dynamic motion of collective secondary electrons, specimen drift occurred owing to a severe charging effect. This indicates that electric circuits comprising various orbits of secondary electrons tend to form at the intensity level of the incident electrons, which is slightly lower than the limit that produces specimen drift. In other words, the formation of electric circuits contributes to stabilization of the specimen, thereby suppressing specimen drift under electron irradiation.

9.1.2 Simulation of Electron Orbits Around Microfibril

Figure 9.5a,b, respectively, show an electron hologram and a reconstructed phase image obtained from microfibrils of sciatic nerve tissue that were fixed and coated with OsO4 and C. The reconstructed phase image shows the electric potential distribution around the charged microfibril. The irregular contrast (indicated by the red

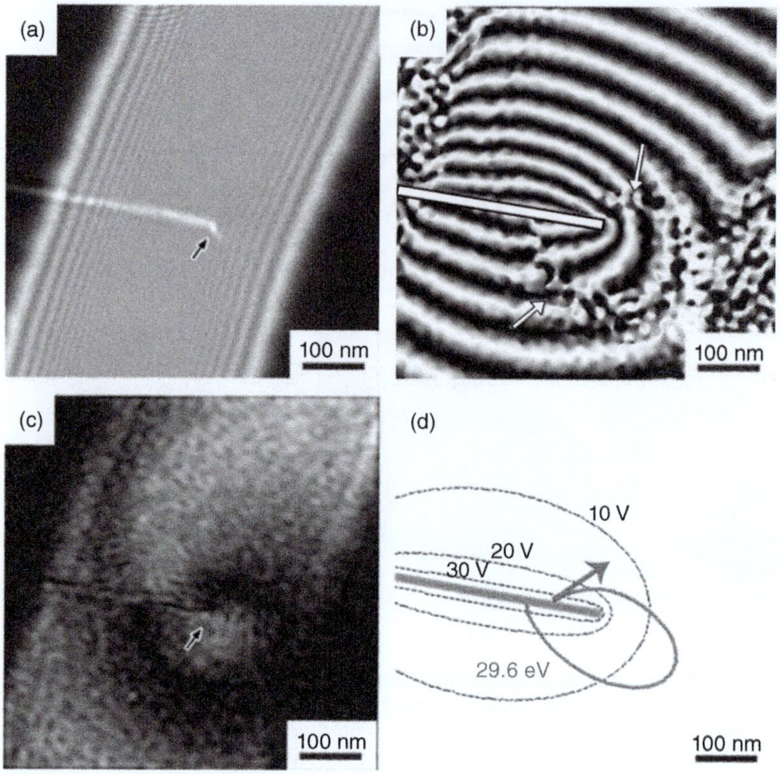

Figure 9.5 (a) Electron hologram of microfibril of sciatic nerve tissue. (b and c) Reconstructed phase and amplitude images, respectively, obtained from hologram. (d) Simulated electron-induced secondary-electron orbits. Contour lines around microfibril indicate electric potential on plane perpendicular to incident electron beam. Source: Inoue et al. (2012), with permission from Oxford University Press.

arrows) corresponds to the electric potential fluctuation and is interpreted to be the stationary orbits of secondary electrons [3]. However, because of the contrast of the black and white bands corresponding to the static electric potential distribution as a result of the specimen being charged, it is difficult to trace the irregular contrast regions.

We applied an amplitude reconstruction process to the hologram shown in Figure 9.5a and obtained the image in (c). The contribution of the static electric potential distribution due to specimen charging is removed, and a simple dark contrast is observed. The dark-contrast region indicates a stationary orbit of secondary electrons around the charged microfibril. The reason of the formation of a stationary orbit of secondary electrons is considered to be the uneven surface of the microfibril as noted above.

Figure 9.5d shows the simulated orbits of the secondary electrons. The parameters for the simulation were obtained by determining the shape of the microfibril from a TEM image and the electric potential of the charged microfibril from the reconstructed phase image (Figure 9.5b). The passing point and angle of the secondary electrons on the surface were also determined from the reconstructed amplitude image. The microfibril was assumed to be cylindrical with a diameter of 15 nm and a length of 360 nm. Its electric potential was set to 38 V. The passing point and angle of the secondary electrons on the surface of the microfibril are indicated by the blue arrow in the figure; the passing point was 75 nm from the tip edge of the microfibril and 1.5 nm above the specimen surface. The angle of the passing direction from the surface plane of the microfibril was 45°. The initial kinetic energy of the secondary electrons at the surface was 29.6 eV. The simulated orbit of the secondary electrons is indicated by the red line with the passage of time 1.0×10^{-15} seconds. It is consistent with the orbit in the reconstructed amplitude image.

Figure 9.6 shows the simulated orbits of secondary electrons as a function of the electric potential at the microfibril surface and the kinetic energy of the electrons. The orbits of the secondary electrons, indicated by red curves, are sensitive to both the electric potential of the specimen surface and their kinetic energy. It can be seen that, at lower energies, the secondary electrons tend to be attracted to the charged specimen while at higher energies they revolve around or leave the charged microfibril. When the electric potential of the microfibril increases, most of the secondary electrons are strongly attracted to it.

So far, we have discussed the orbits of the secondary electrons outside the microfibril. We now consider the motion of these electrons on its surface. There is a sharp black line corresponding to a part of the orbit of the secondary electrons on the microfibril, as indicated by an arrow in Figure 9.5c. To clarify the situation, we superimposed the hologram on the reconstructed amplitude image, as shown in Figure 9.7a. The linear orbit of the secondary electrons along the microfibril between the blue arrowheads is clearly evident.

According to the simulation results shown in Figure 9.6, most of the secondary electrons with lower energies are attracted to the specimen surface. Since a microfibril fixed with OsO4 and coated with C still has an insulating state, some of the lower-energy secondary electrons exist on the surface of the charged microfibril. Eventually, secondary electrons far from the positively charged microfibril are attracted to the specimen, but when they approach its surface, they suffer a repulsive force (according to the inverse square law) due to the Coulomb force from the negative charges on the specimen. To support this interpretation, a detailed electric field analysis around the surface of the charged specimen is necessary. We therefore performed a simulation with a simple model, i.e. a model with a single layer consisting of negative point charges (electrons) assumed to have lower energies situated 0.1 nm above the specimen. The charges were arranged with a plane density of $1.0\,\text{nm}^{-2}$, as shown in Figure 9.7b,c. The red dot and red arrow in (c), respectively, indicate the initial point and the angle of the secondary electrons starting from a position 1.4 nm above the charged layer. The passage of time for the calculation of the orbits of the secondary electrons was 1.0×10^{-17} seconds. Portions of the simulated orbits of secondary electrons with initial kinetic energies of 25 and 30 eV at the initial point are represented in Figure 9.7d by purple and red curves, respectively. The electrons approach the surface because of the attractive Coulomb force of the positively charged microfibril. However, they do not reach it because of

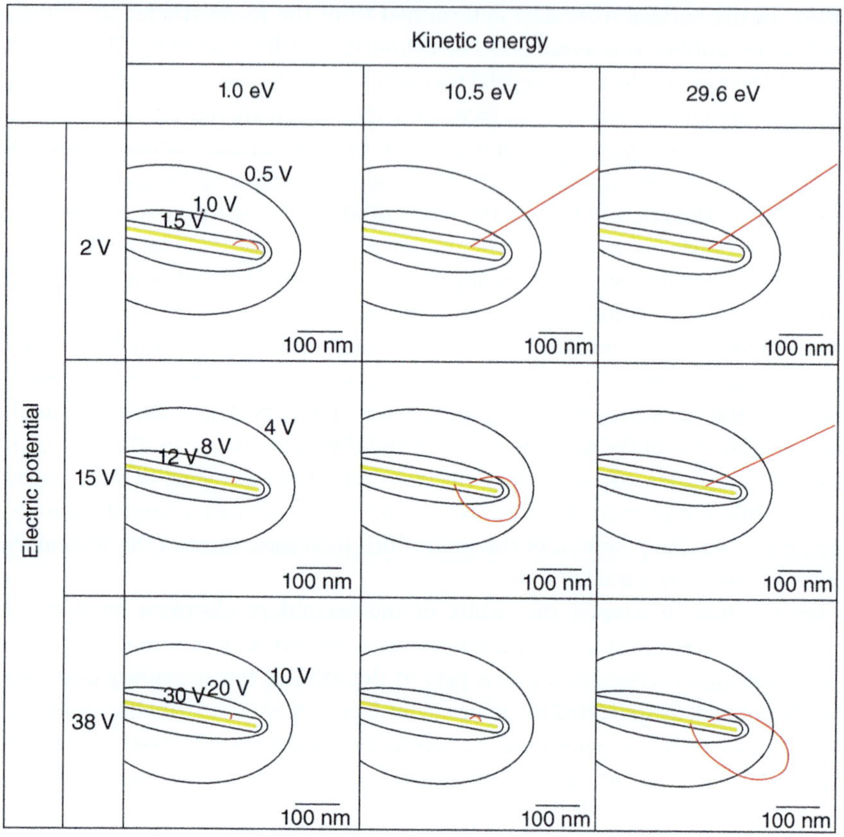

Figure 9.6 Simulated orbits (red lines) of secondary electrons as function of electric potential of charged microfibril and electron kinetic energies.

the strong repulsive force with the inverse square law due to the electron layer on the surface. They tend to glide a few nm above the surface, as shown in the figure.

From the discussion [5] and the simulation results described here, the mechanism for the formation of electric circuits can be understood as follows. Microfibrils tend to be positively charged, particularly at the top-edge regions (see Figure 9.2), because of the emission of secondary electrons. This feature can be seen in the slightly bright region indicated by thin black arrows in the holograms and the reconstructed amplitude images in Figures 9.5a,c and 9.7a.

Although secondary electrons are initially emitted in various directions with various energies, stationary orbits are considered to form depending on the surface morphology and the charged state of the specimen. Secondary electrons emitted with relatively high energies tend to revolve spherically around the top-edge regions. When they approach the uneven surface of the microfibril body, they encounter a repulsive force due to the presence of lower-energy secondary electrons around the surface and tend to glide along the surface, being gradually pulled toward the strongly charged top-edge region. Around the top-edge region, the

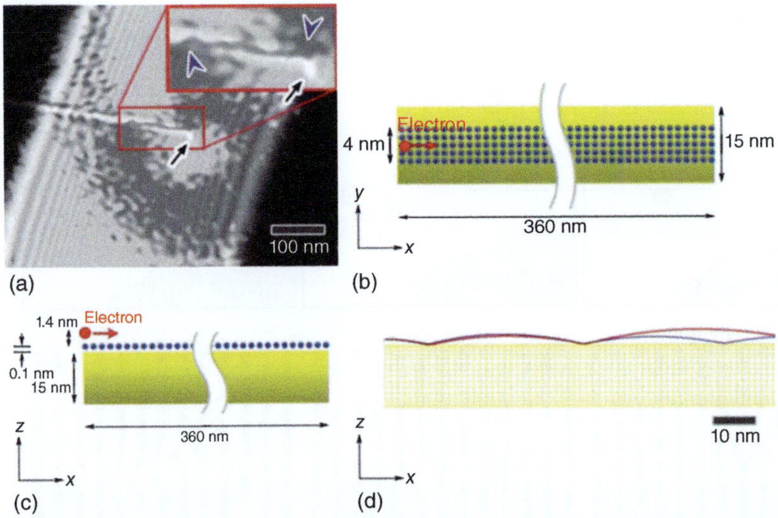

Figure 9.7 (a) Hologram superimposed on its reconstructed amplitude image. Red rectangle is enlarged in inset. (b, c) Simple model for simulation indicating distribution of electrons on positively charged specimen surface. Directions of incident electron beam and line of microfibril are parallel to *z*-axis and *x*-axis, respectively. (d) Purple and red curves correspond to orbits of secondary electrons on surface of microfibril with kinetic energies of 25 and 30 eV, respectively. Source: (a) Inoue et al. (2012), with permission from Oxford University Press.

secondary electrons leave the microfibril surface with the kinetic energy and tend to form electric circuits. This behavior of secondary electrons leaving the microfibril surface and forming electric circuits is attributable to the surface's characteristic morphology and to the inhomogeneous charge distribution [3].

9.1.3 Interpretation of Reconstructed Amplitude Image

Figure 9.8 schematically illustrates the reconstruction process of an amplitude image, including the electron motion. The velocities of indent electrons with energies of a few hundred keV are about 100 times higher than those of secondary electrons with energies of a few dozen eV. Therefore, the first incident electron is influenced by the electrostatic potential due to the charge distribution (Figure 9.8a-1). After the first electron passes through the specimen, the following incident electrons are influenced by the electric potential due to the charge distributions ((a-2) and (a-3)). Note that the electric potential at the insulating specimen position in the lower left is basically stationary [6], while the electric potential along the path of the secondary electrons continually changes. Thus, the resulting interference fringes tend to change gradually along the path of the secondary electrons, as schematically shown in Figure 9.8b-1–b-3.

What we observe is the average of the interference fringes recorded during the exposure time of about several seconds, as shown in Figure 9.8c-1, during which the fringe contrast along the path of the secondary electrons decreases. The digital

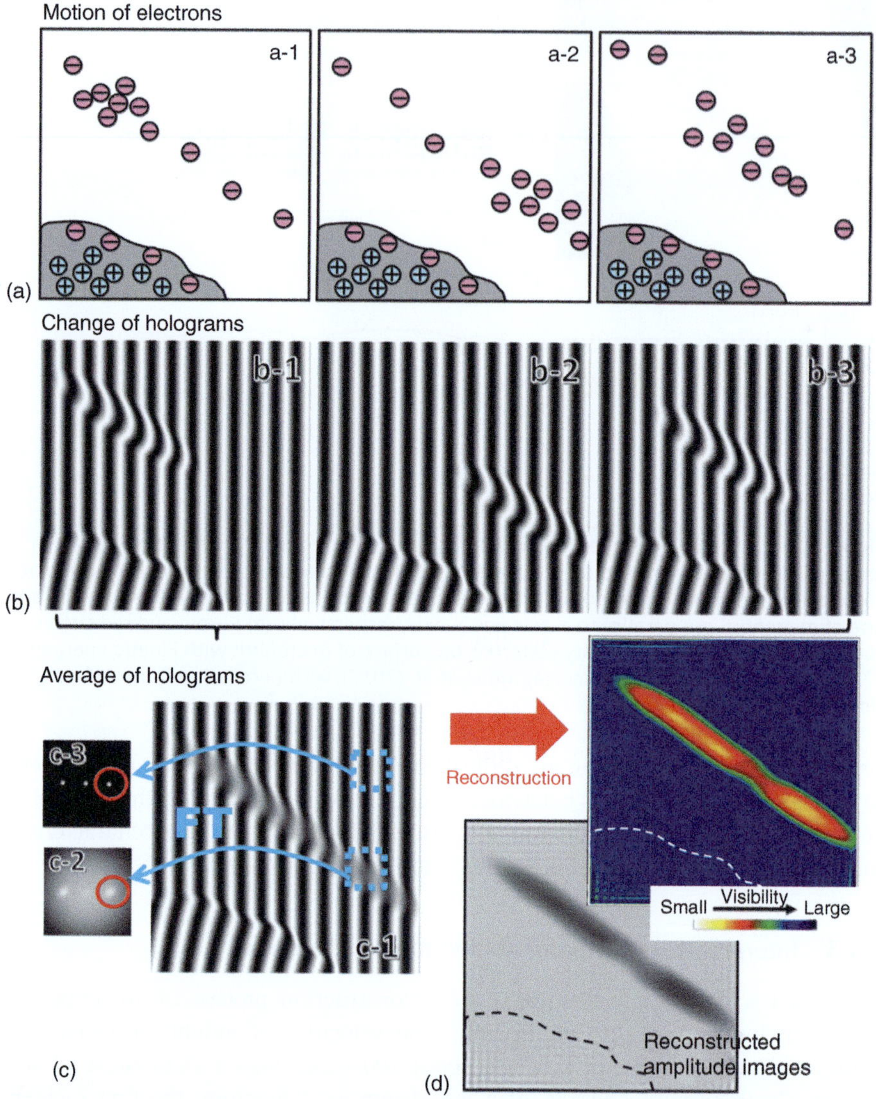

Figure 9.8 Schematic illustration of amplitude reconstruction process for electron motions. (a) Distributions of electrons around charged specimen. (b) Change in interference fringes due to electron motions. (c) Contrast of averaged holograms (c-1). Note high amplitudes of background in digital diffractogram (c-2) obtained from region around specimen surface through Fourier transformation. (d) Amplitude image obtained from hologram (c-1). Dotted lines in (d) indicate specimen edge. Source: (c) Shindo and Akase (2020), with permission from ELSEVIER.

diffractograms obtained from the path of the secondary electrons and from another area around the specimen surface are shown in Figure 9.8c-2 and c-3, respectively. Note that the amplitudes of the background in (c-2) are large. In the reconstruction process, a sideband region is selected and processed using an inverse Fourier transformation. Since the amplitudes outside the selected region are missing, a dark contrast appears along the path of the secondary electrons, as shown in Figure 9.8d. It is also noted that the contrast of reconstructed amplitude images along the path of the secondary electrons is directly related to the visibility of interference fringes. To make clear the contrast change of reconstructed amplitude images due to the electric field variations, color images with the color scale indicating the relation between the visibility interference fringes and the contrast of the reconstructed amplitude images are utilized as shown in (d) in the following sections.

9.1.4 Simulation of Visibility of Interference Fringes for Electron Motion

The contrast of reconstructed amplitude images along the path of the secondary electrons is directly related to the visibility of interference fringes, which is also related to the fluctuation in the average density of secondary electrons projected along the incident electron beam as noted above. To clarify these relationships quantitatively, the visibility of interference fringes for electron motion was simulated to clarify the effect of the electric potential due to secondary electrons on the phase shift of incident electrons or on the distortion of the interference fringes [3]. Holograms were simulated on the basis of region S in Figure 9.9a, in which the secondary electrons were simply assumed to stand still, since the velocity of incident 300-keV electrons ($v = 2.33 \times 10^8\ \mathrm{m\,s^{-1}}$) is considerably higher than that of secondary electrons with energies ranging from a few to a few tens of eV (v: order of magnitude $1 \times 10^6\ \mathrm{m\,s^{-1}}$).

Given the spherical distribution of the orbits of secondary electrons, we consider the density of secondary electrons projected along the incident electron beam to be higher along the great circle corresponding to region S1 in Figure 9.9a. When the electrons are relatively few in number, e.g. 33, 15, 10, and 3, and are assumed to occupy random positions in regions S1 ($30\ \mathrm{nm} \times 2\ \mathrm{nm}$), S2 ($30\ \mathrm{nm} \times 2\ \mathrm{nm}$), S3 ($30\ \mathrm{nm} \times 4\ \mathrm{nm}$), and S4 ($30\ \mathrm{nm} \times 6\ \mathrm{nm}$), respectively, a small modulation in interference fringes is observed, as shown in (b), where the black and white bands along the y-axis have 2-nm separation and the sinusoidal intensity distribution corresponds to the interference fringes in the hologram.

When the number of electrons is twice that in the case shown in Figure 9.9b, which corresponds to an electric current of 1 mA when the velocity of the secondary electrons is assumed to be $1.5 \times 10^6\ \mathrm{m\,s^{-1}}$, considerable distortions of the interference fringes are observed, as shown in (c). We attribute the observed disappearance of interference fringes to the inhomogeneous current of secondary electrons in the orbits around and along the microfibrils; interference fringes tend to disappear

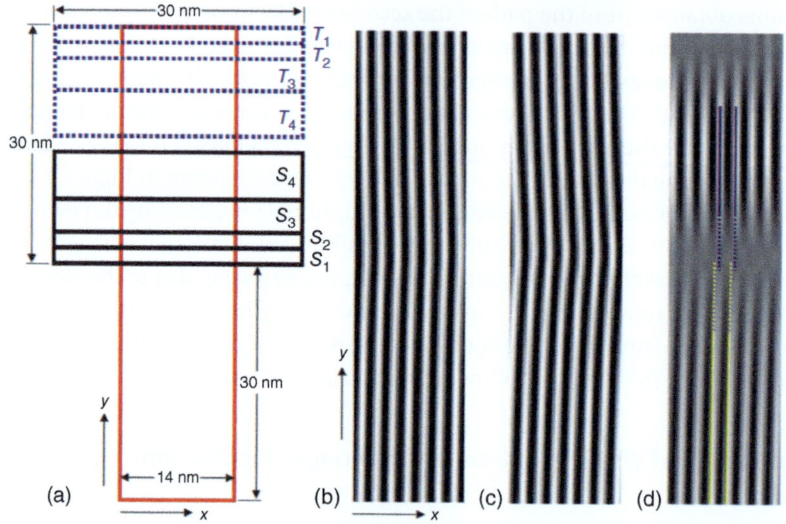

Figure 9.9 (a) Simple model for part of electron orbits projected along incident electron beam. Region S (outlined with solid lines and containing secondary electrons) is assumed to simulate holograms in (b) and (c) while regions S and T are both considered to simulate hologram in (d). Areas of holograms in (b)–(d) correspond to red rectangle in (a). In (b), projected densities of secondary electrons in S1, S2, S3, and S4 are 0.55, 0.25, 0.083, and 0.017 nm^{-2}, respectively. Twice the densities in (b) is assumed in (c). The hologram in (d) was obtained assuming the same densities as those in region S in (c) for region S or T.

owing to the blurring effect if the fluctuation in the phase change of the incident electrons is around π.

To see this blurring effect, another region, i.e. region T, with the same size and electron density as region S is assumed, as indicated by the dotted lines in Figure 9.9a. A hologram is obtained by averaging the intensities of the interference fringes in two states in which secondary electrons exist in region S or T. Note that the number of electrons or the total electric charge is the same for the two states and that the average electric potential in regions S and T is lower than that in the outside region (lower part of figure).

The hologram in Figure 9.9d shows the disappearance of interference fringes in region S1, which corresponds well to the image contrasts in Figure 9.3b. The disappearance of interference fringes is also observed in region T1 in Figure 9.9d. Furthermore, in the simulated image in Figure 9.9d, as indicated by the dotted yellow lines, extrapolated lines along the interference fringes shift away from each other, which is consistent with the shift in region E in Figure 9.3b.

Here, the visibility of the interference fringes, V, is given by

$$V = (I_{\max} - I_{\min})/(I_{\max} + I_{\min}) \tag{9.1}$$

where $I_{\max}$ and $I_{\min}$ are the maximum and minimum intensities in each region. This visibility is directly related to the contrast of reconstructed amplitude images. In this way, the color images with the color scale can be presented as shown in

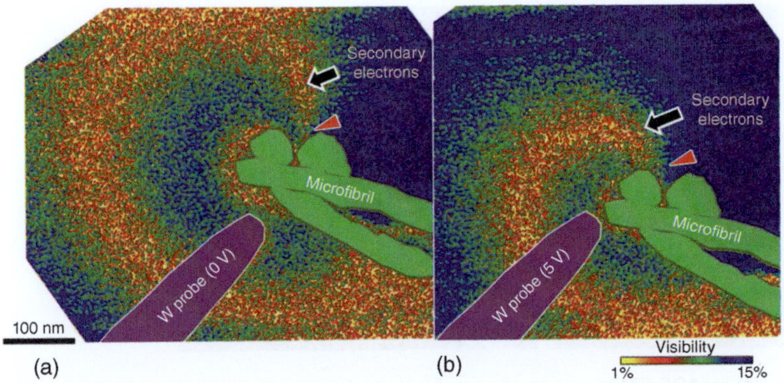

Figure 9.10 Reconstructed amplitude images showing change in electric circuits of secondary electrons resulting from insertion of electrode. Color bar indicates visibility of fringe contrast in original hologram. (a) Electrode (tungsten) outlined in white; bias voltage of electrode is zero. (b) Bias voltage of electrode is +5 V.

Figure 9.8d. Thus, the reconstructed amplitude images with color scale are utilized for the motion and distribution of secondary electrons in the following sections.

9.1.5 Change in Electron Orbits Due to Insertion of Electrode

As shown in Figure 9.10a, when a tiny electrode is inserted near the microfibrils, the small electric circuit around the top region of the microfibrils and the circuits around region D in Figure 9.3d disappear while a few simple circuits are left because of the change in the electric field. Furthermore, when a bias voltage of 5 V is applied to the electrode, the circuit becomes smaller, as shown in Figure 9.10b. Although the sizes and shapes of the electric circuits change due to the field produced by the inserted electrode, one of the spreading regions of secondary electrons is located at the same position, as indicated by the red arrowheads.

While the electric circuits that form around tangled microfibrils are stable, the electric circuit that forms around a single microfibril is not so stable, and its position and size gradually change [3]. It has been observed that the surface of a microfibril is uneven and that several dark spots, which are considered Os-rich regions, appear [3]. High-resolution high-angle annular dark field (HAADF) STEM has been used to observe these Os-rich regions shown as white dots in Figure 9.11a. The appearance of these regions has been also confirmed by energy-dispersive X-ray spectroscopy (EDS) elemental mapping image, as shown in (b). The uneven surface and inhomogeneous distribution of the coating element apparently cause an inhomogeneous charge distribution on the surface of the microfibrils, resulting in the characteristic motions of collective secondary electrons. These features were clarified well by simulating the orbits of secondary electrons as functions of the electric potential of a charged microfibril and the kinetic energy of the secondary electron as noted in Section 9.1.2. Sections 10.1, 10.2, and 11.2 further discuss the effects of metallic elements present on various insulating material surfaces on the accumulation and motion of secondary electrons.

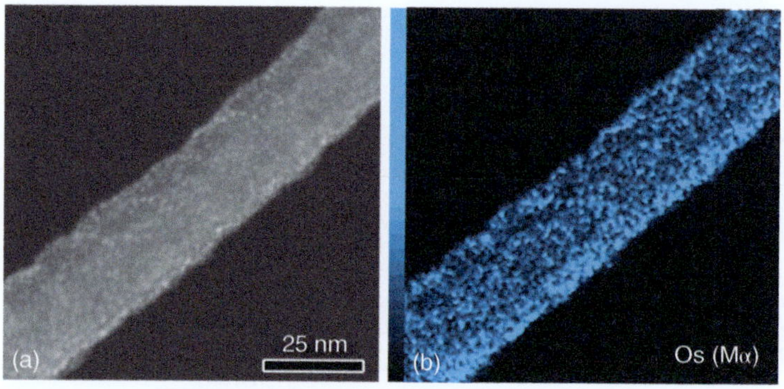

Figure 9.11 (a) HAADF-STEM image of microfibril. (b) Os (Mα) mapping image of microfibril obtained by EDS. Images were obtained in collaboration with Mr. A. Yasuhara. Source: Shindo and Akase (2020), with permission from ELSEVIER.

9.2 Visualization of Accumulative and Collective Motions of Electrons

Figure 9.12a shows a TEM image of a microfibril with uneven surfaces ending in two branches. The hologram in (b) corresponds to the top right region, indicated by the white square in (a) [7]. The incident electron intensity was about $0.15\ \text{nm}^{-2}\cdot\text{s}^{-1}$, much smaller than that used in the studies on visualizing electron orbits discussed above.

Figure 9.13a shows a reconstructed amplitude image observed at the initial state while (b) and (c) show reconstructed amplitude images with the passage of time [7]. The bright red regions, which are considered to correspond to the fluctuation of the electric potential due to the motion of the secondary electrons, gradually shift between the branches. Contrast formation in reconstructed amplitude images due

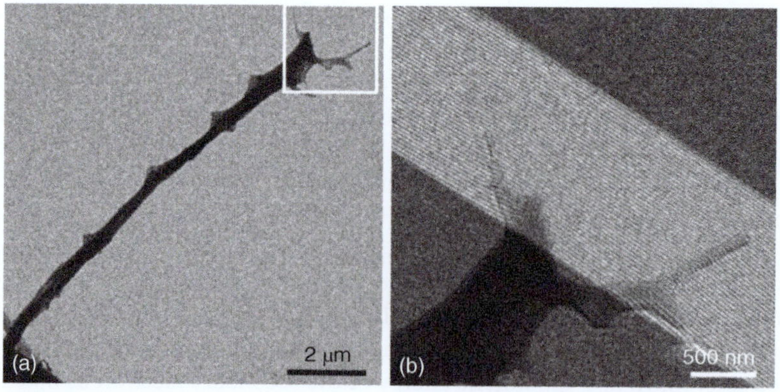

Figure 9.12 (a) Electron micrograph and (b) hologram of microfibril of sciatic nerve tissue observed at initial state. Source: Shindo et al. [7], with permission from Cambridge University Press.

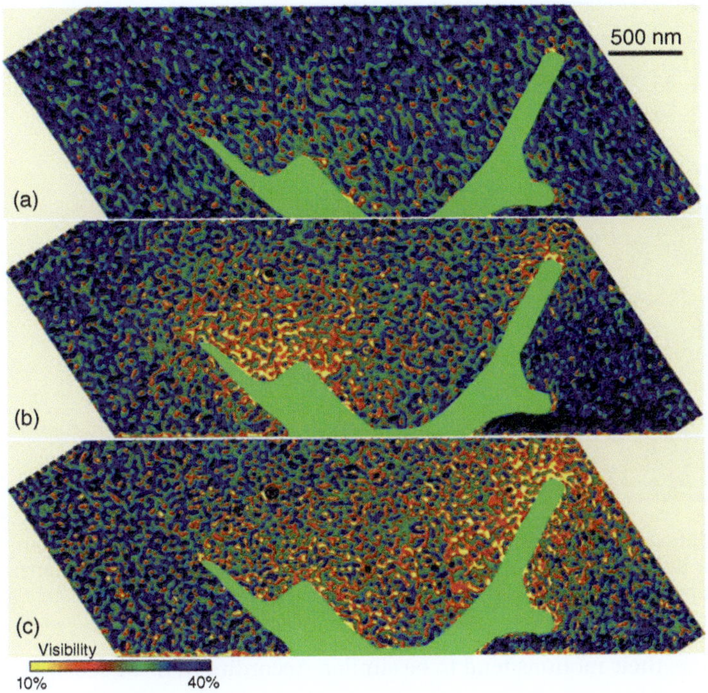

Figure 9.13 (a) Reconstructed amplitude images obtained from hologram of microfibril at initial state shown in Figure 9.12b. (b and c) Amplitude images reconstructed from holograms recorded with passage of time from initial state.

to the accumulation of electrons around charged microfibrils was also shown in a similar manner to Figure 9.8 [4].

The motions of electrons confined in the branch region were also explained through a simulation taking into account the interaction of the electrons and the surfaces of the charged specimen. Figure 9.14 shows the results of a simulation performed to clarify the motion of secondary electrons. The simulation was carried out simply in two dimensions, i.e. in a plane including the two branches of a microfibril. The electrostatic potential at the specimen surface was assumed to be 15 V on the basis of simulated data for the average reconstructed phase image [7]. When the initial energy of the secondary electrons was low (3 eV), secondary electrons #1 and #2 (emitted from both sides of the left branch) returned to the surface of the same branch. When secondary electron #3 with an initial energy of 10 eV was emitted from the left-side surface of the left branch, however, it moved downward and did not return. In contrast, when secondary electron #4 was emitted from the right-side surface of the same branch, it moved close to the surface of the right branch because of the branch's attractive Coulomb force. Since there are at least several electrons on the surface, such as secondary electrons with lower energy, secondary electron #4 did not reach the surface because of the strong repulsive Coulomb force of the electron layer on the surface [8]. It tended to move back toward its origin on the left branch. When secondary electrons are emitted

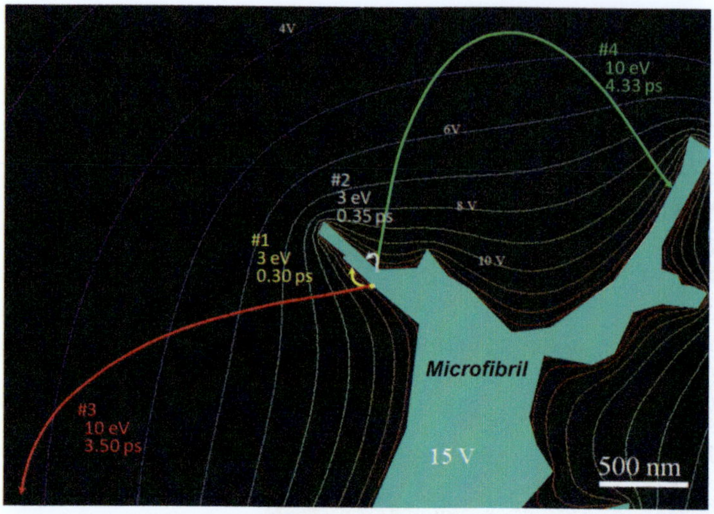

Figure 9.14 Simulation results showing trajectories of secondary electrons emitted from left branch. Contour lines indicate electric potential at and around microfibril in the plane. Initial energy and time of flight are given for each secondary electron.

slightly out of plane, their motions tend to be similar. Accordingly, these secondary electrons are confined between the two branches, as shown in Figure 9.14.

References

1 Shindo, D., Kim, J.J., Xia, W. et al. (2007). *J. Electron Microsc.* 56: 1.
2 Kim, K.H., Akase, Z., Shindo, D. et al. (2013). *Microsc. Microanal.* 19 (Suppl 5): 54.
3 Shindo, D., Jung Kim, J., Hyun Kim, K. et al. (2009). *J. Phys. Soc. Jpn.* 78: 104802.
4 Shindo, D., Tanigaki, T., and Park, H.S. (2017). *Adv. Mater.* 29: 1602216.
5 Cowley, J.M. (1990). *Diffraction Physics*. New York: North-Holland.
6 Sato, T., Tsukida, N., Higo, M. et al. (2019). *Mater. Trans.* 60: 2114.
7 Shindo, D., Aizawa, S., Akase, Z. et al. (2014). *Microsc. Microanal.* 20: 1015.
8 Inoue, M., Suzuki, S., Akase, Z., and Shindo, D. (2012). *J. Electron Microsc.* 61: 217.

10

Collective Motions of Electrons Around Various Charged Insulators

In the observations of the collective motions of electrons around charged biological specimens described above, the distribution and motions of the electrons directly depended on the shape and morphology of the microfibrils of sciatic nerve tissue. Because electromagnetic fields generally originate from various motions of electrons in many devices, it is of vital importance, as well as of particular interest, to control and visualize the collective motions of electrons with various insulators for which the shapes can be controlled using an FIB method and/or ultramicrotomy. In the following sections, we present studies to control and visualize the collective motions of electrons with various insulating materials in comparison with observations of microfibrils of sciatic nerve tissue.

10.1 Accumulation of Electrons on Cleaved Surfaces of $BaTiO_3$

Ferroelectric domain walls [1] and irradiation damage [2] in $BaTiO_3$ have been studied using electron holography. In particular, the electric field variations occurring with temperature change around the ferroelectric Curie temperature [3, 4] and the nanoscale polarization induced by applying an electrical bias [4] were studied. The electric potential distribution in model capacitors consisting of single-crystalline $BaTiO_3$ and thin electrode plates made of Pt was quantitatively analyzed by utilizing a double-probe piezodriving holder [5]. In situ electric field observation of small precipitates in $BaTiO_3$ multilayer ceramic capacitors has also been carried out [6].

With $BaTiO_3$ utilizing its cleavage property, we investigated the charging effect and motions of secondary electrons. First, we simply attempted to reproduce the result obtained for a microfibril with two branches, i.e. the accumulation of secondary electrons between the branches, by using an FIB method for controlling the specimen shape.

Figure 10.1a shows a scanning ion microscopy (SIM) image of a $BaTiO_3$ specimen with two branches at the top. The thickness of the specimen along the incident electron beam direction was 6 μm. Figure 10.1b shows a reconstructed phase image of the

Material Characterization using Electron Holography, First Edition. Daisuke Shindo and Takeshi Tomita.

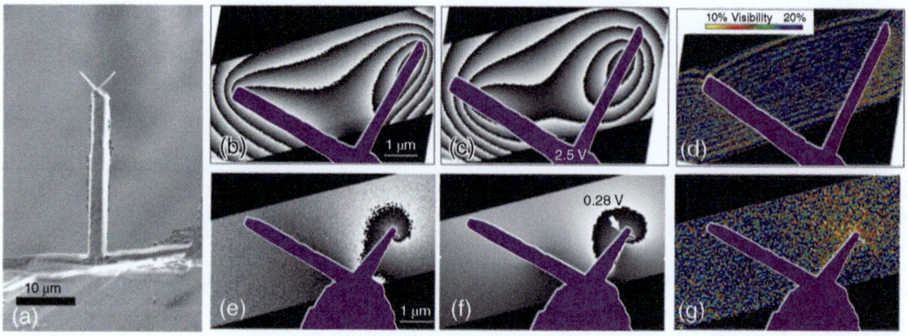

Figure 10.1 Charging effect and secondary electron distribution of $BaTiO_3$. (a) SIM image showing branch region of $BaTiO_3$. (b) Reconstructed phase image of $BaTiO_3$. (c) Simulated phase image of same region. Electric potential of surface was assumed to be 2.5 V. (d) Reconstructed amplitude image of $BaTiO_3$. (e) Reconstructed phase image of specimen with cleaved edge in right branch. (f) Simulated phase image of same region. Electric potential of surface was assumed to be 0.1 V, except at cleaved edge, where it was assumed to be 0.28 V. (g) Reconstructed amplitude image of specimen with cleaved edge. Source: Shindo and Akase [7], with permission from ELSEVIER.

top-edge region in (a). The branches are shown in purple. From the simulation phase image shown in (c), the electric potential of the specimen was estimated to be 2.5 V. In the reconstructed amplitude image shown in (d), unlike the region described above in which the microfibril of sciatic nerve tissue was branched, secondary electron accumulation was not observed between the branches. The faint contrast, visible below the right branch, suggests only a modest degree of accumulation.

This difference in the electron distribution is attributed to the formation of a conductive layer on the specimen surface during FIB thinning, which was clarified by electron probe microanalysis (EPMA). After successive TEM observation and EPMA, part of the right branch was broken by mechanical shock during the treatment and transfer of the specimen, as evident in Figure 10.1e. The electric potential in most of the branch region was low (0.1 V) due to substantial carbon contamination and the low intensity of the incident electron beam. The fresh surface edge formed by the cleavage, however, has a localized contour line that corresponds to an electric potential of 0.28 V, as seen in the simulation result in (f). The reconstructed amplitude image in (g) clearly shows bright colored regions around the fresh surface area, which indicates that secondary electrons with a fluctuating electric field tend to accumulate around the fresh surface of a charged specimen [8]. These results demonstrate that the collective motions of electrons are sensitive to the surface conditions of insulating materials with or without conductive materials on the surfaces. This feature can be utilized for controlling the accumulation of electrons around insulating materials, as explained below.

Taking these results into account, we artificially controlled the shape of the specimen to form a localized fresh surface devoid of conductive materials for clearer visualization of the secondary electron distribution. Figure 10.2a shows an SIM image of a $BaTiO_3$ rod (triangular prism shape) prepared using an FIB method. The red arrow indicates the thin neck region formed at the wedge tip. Immediately after

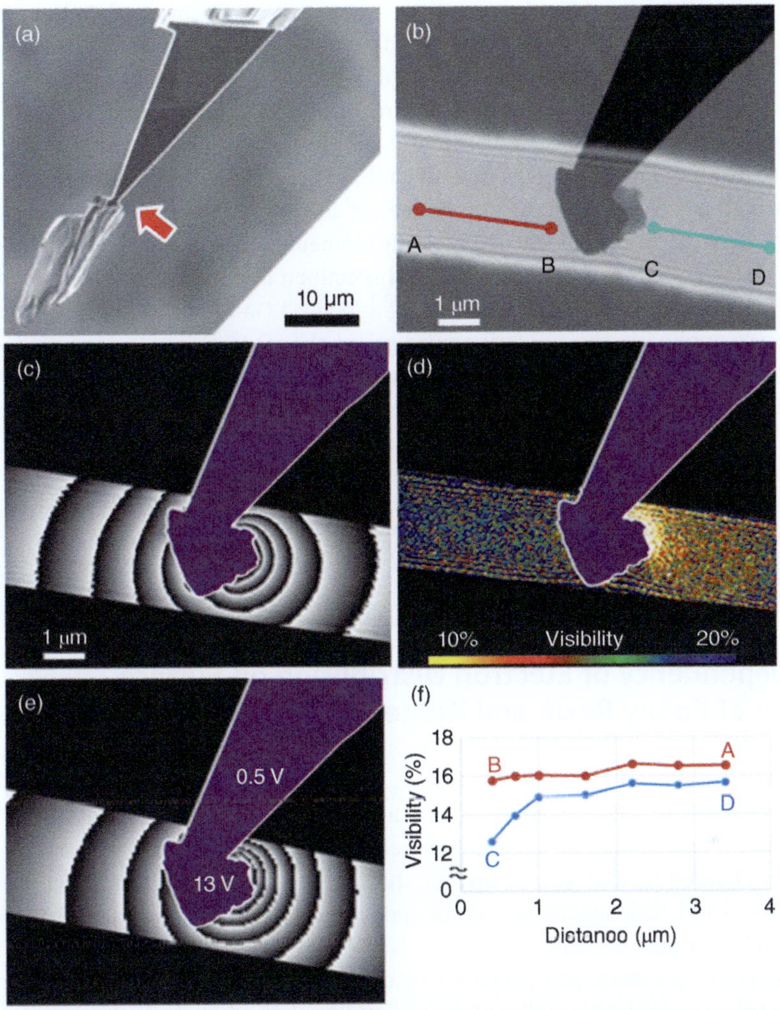

Figure 10.2 Distribution of secondary electrons around cleaved $BaTiO_3$ rod tip. (a) SIM image of $BaTiO_3$ rod before cleavage. (b) Hologram of $BaTiO_3$ rod tip after cleavage. (c) Reconstructed phase image of $BaTiO_3$ rod tip obtained from hologram. (d) Reconstructed amplitude image obtained from hologram. (e) Simulated phase image corresponding to (c). (f) Visibility as function of distance from specimen between points indicated in (b). Source: Shindo and Akase [7], with permission from ELSEVIER.

FIB processing, the specimen was cleaved by applying a mechanical shock with a glass fiber. The cleaved specimen was then mounted on a specimen holder.

Figure 10.2b shows a hologram of the upper region of the cleaved specimen. The reconstructed phase image in (c) shows that the electric potential reached a high value (about 13 V) in the cleaved region with the fresh surface from the simulated phase image shown in (e). In the simulation, the specimen thickness in the cleaved area was set to 200 nm, the thickness of the other regions was set to 2.9 μm, and the electric potential for the surface was set to 0.5 V. In the reconstructed amplitude image in Figure 10.2d, bright red and yellow speckles are evident on the cleaved surface and extend over a wide area. The variations in visibility in the hologram measured along paths B–A and C–D (see Figure 10.2b) are compared in (f). The visibility is minimal in region C, i.e. near the freshly cleaved surface, and increases toward region D while remaining lower than that between B and A [8].

The above results shown in Figure 10.2 mean that a secondary electron manipulator can be developed by utilizing the phenomena of accumulation of secondary electrons being sensitive to the existence of metallic elements on the surface of charged insulating materials. An insulating $BaTiO_3$ rod specimen like the one shown in the SIM image in Figure 10.2a was prepared using a focused Ga-ion beam. A mechanical shock was applied to the rod (triangular prism shape) to cleave the specimen in the thin neck region at the wedge tip, resulting in the formation of a localized clean surface. Observation of this specimen by electron holography revealed that secondary electrons accumulated locally around the clean surface region, as shown in (d). Mounting of this specimen in a specimen holder with a piezodriving probe (Figure 6.28) and moving to semiconductor specimens enable direct observation of the interactions of the secondary electrons with the specimen surfaces and of various internal defects.

10.2 Dependency of Electron Distribution on Surface Condition of Epoxy Resin and Kidney

In the work described in the previous section, the collective motions of electrons were controlled by using Ga^+ irradiation with an FIB system and the cleavage property of $BaTiO_3$. The dependency of secondary electron accumulation on the surface condition of charged insulators was further studied by combining ultramicrotomy and an FIB method. The specimen was an epoxy resin, a rather soft material that can be sliced by ultramicrotomy.

First, an epoxy specimen was prepared by ultramicrotomy, resulting in a clear specimen surface devoid of metallic elements, unlike that when an FIB system is used. Figure 10.3a shows a reconstructed phase image of a thin film of epoxy resin (dark brown region) prepared by ultramicrotomy. To estimate the electric potential of the specimen's top surface, a reconstructed phase image was simulated. The thickness of the model was 2.0 μm, and the electric potentials on the sides and bottom of its surfaces were set to 0.0 V. The width of the interference fringe region was assumed to be 2.8 μm. Compared with the simulation results shown in (b), the electric potential of the top surface of the epoxy resin was estimated to be 1.2 V. Although there were no metallic elements on the surface of the specimen, the charging effect is considered to have been partly suppressed by the irradiation of secondary electrons from the specimen support plate near the observed area.

The reconstructed amplitude image in Figure 10.3c shows red regions around the surface of the epoxy resin; they are considered to correspond to high densities of secondary electrons strongly interacting with the surface of the positively charged

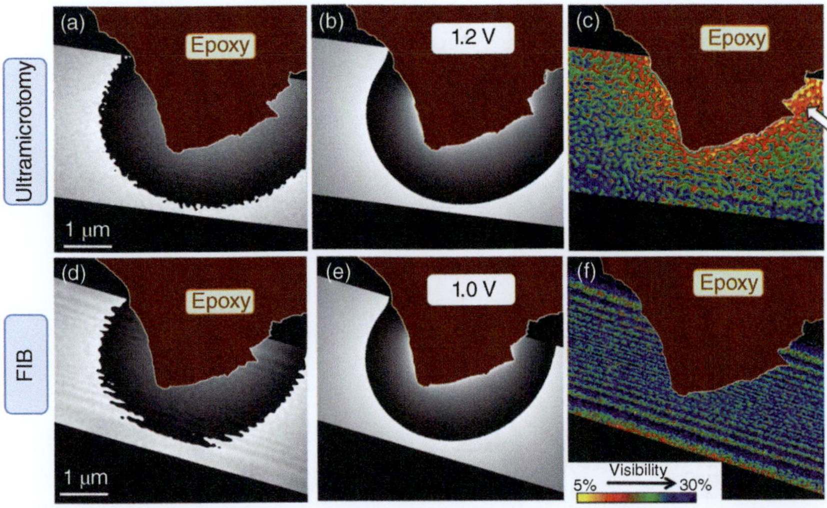

Figure 10.3 (a) Reconstructed phase image of ultramicrotomed epoxy resin. (b) Simulated phase image with surface electric potential of 1.2 V. (c) Reconstructed amplitude image of ultramicrotomed epoxy resin. (d) Reconstructed phase image of epoxy resin treated with FIB system. (e) Simulated phase image with electric potential of 1.0 V. (f) Reconstructed amplitude image of epoxy resin treated with FIB.

specimen. Particularly in the concave region, indicated by the arrow in the figure, a bright red region is evident. It corresponds to a large electric field fluctuation due to interaction of the accumulated secondary electrons with the surface, similar to the case of the branched microfibril in Figure 9.14.

After observation of the hologram of the epoxy resin, both sides of the thin specimen were irradiated using an FIB system and a weak Ga-ion beam. The beam intensity was 0.85×10^3 mC m^{-2}, which is 200 times less than that typically used for polishing specimens. The reconstructed phase image in Figure 10.3d shows that the electric potential of the specimen was 1.0 V in accordance with the simulated phase image shown in (e). The electric potential of the specimen before and after irradiation did not differ significantly because of the weak Ga-ion beam. Note that the shape of the specimen in (d) is almost the same as that in (a), which indicates that the irradiation damage caused by the Ga ions was negligible.

In the reconstructed amplitude image in Figure 10.3f, the red regions evident in (c) are no longer visible. Thus, the distribution of secondary electrons that interact strongly with the surface of positively charged epoxy resin is quite sensitive to the presence of metallic elements on the surface [9].

Similarly, a thin biological specimen (a flake of mouse kidney tissue) was prepared by ultramicrotomy, and the dependency of the secondary electrons on the surface conditions was investigated and compared with that of epoxy resin. Figure 10.4a shows a reconstructed phase image of the specimen embedded in resin (blue color). From the simulation of the reconstructed phase image (b), the electric potential of the kidney flake was estimated to be 3.2 V.

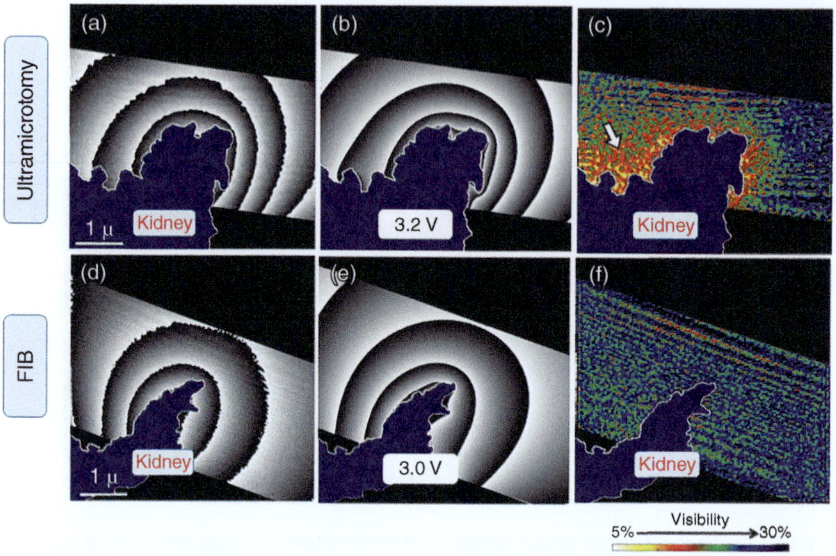

Figure 10.4 (a) Reconstructed phase image of kidney flake prepared by microtomy. (b) Simulated phase image of kidney flake prepared by microtomy when surface electron potential was estimated to be 3.2 V. (c) Reconstructed amplitude image of kidney flake prepared by ultramicrotomy. (d) Reconstructed phase image of kidney flake treated with FIB system. (e) Simulated phase image of kidney flake treated with FIB system when surface electron potential was estimated to be 3.0 V. (f) Reconstructed amplitude image of kidney flake treated with FIB method. Source: Image data were obtained in collaboration with Prof. N. Ohno.

The reconstructed amplitude image in Figure 10.4c shows brightly red regions around the surface of the kidney flake. These regions correspond to high densities of secondary electrons. In the concave region indicated by the arrow, the brightly colored regions extend more widely than those on the right of the flake. This is consistent with the above result of the thin film of epoxy resin.

We evaluated the visibility near the surface region to estimate the density of the secondary electrons projected along the incident electron beam and estimated it to be approximately 0.5 nm^{-2}. After observing the hologram of the kidney flake specimen, we irradiated both sides of it with Ga ions over a period of eight hours. The intensity of the Ga-ion beam was set to 0.85×10^{-3} mC m^{-2}, which (as noted above) is 200 times less than that usually used for polishing specimens.

The reconstructed phase image in Figure 10.4d shows that the specimen had become smaller, which was due to the Ga-ion irradiation. From the simulation of the reconstructed phase image (e), the electric potential of the specimen was estimated to be 3.0 V, which is slightly less than that observed before the Ga-ion irradiation. The brightly red regions observed in (c) are not visible in the reconstructed amplitude image in (f). The absence of secondary electrons around the surface is considered to result from the transfer of electrons through the conductive layer on the surface formed by the Ga-ion irradiation.

10.3 Electron Distribution Between Epoxy Resin and Kidney

On the basis of these studies on the distribution of secondary electrons on charged insulating films, we performed in situ experiments to investigate the interaction of insulating specimens. Figure 10.5 shows the results of an in situ study of the electric field variations and the change in the secondary electron distribution between two insulating materials: a square pillar of epoxy resin prepared using an FIB method with Ga^+ ions and a flake of kidney tissue embedded in epoxy resin prepared by ultramicrotomy [10]. The geometric configuration of the square pillar and kidney flake is shown in Figure 10.5c. The pillar was placed in the piezodriving probe of a TEM specimen holder and gradually moved toward the kidney flake.

In the reconstructed phase images (Figure 10.5a), the electric potential of both specimens decreased as their separation decreased due to the mutual irradiation with secondary electrons. In the reconstructed amplitude images (b), red regions are evident around the kidney flake but not around the square pillar, which is consistent with the observations presented in Figure 10.3. When the separation of the specimens fell below approximately 1 μm, most of the red regions disappeared, indicating a decrease in the number of secondary electrons around the kidney flake. Nonetheless, faintly colored regions remained in the lower-right region, some distance from the pillar of epoxy resin.

The visibilities of the interference fringes in the green (f) and yellow (e) square regions indicated in the leftmost image of Figure 10.5a are compared in (d) as functions of the separation between the specimens as the pillar moved toward or away from the kidney flake. A prominent increase in visibility was evident in the yellow square near the kidney flake when the pillar approached the kidney flake surface. The decrease in the number of secondary electrons on the surface of the kidney flake was due to the presence of a conductive layer on the surface of the epoxy resin, through which secondary electrons were transferred [10].

10.4 Control of Electron Distribution Around Cellulose Nanofibers by Applying External Electric Field

The collective motions around microfibrils of sciatic nerve tissue treated with OsO_4 are described in Chapter 9. To compare the results with those for microfibrils, we investigated the motions of electrons around a nanofiber specimen devoid of metal elements on the surface. Here, we present the results obtained for cellulose nanofiber (CNF), a promising biomass material made of highly refined wooden fiber with several attractive features. Paper made of CNF has many desirable properties such as high strength, low thermal expansion, and a highly insulative nature, in addition to its high transparency and light weight [11–13]. Various useful properties (e.g. conductivity and electrical capacitance) can be imparted to a sheet of CNF

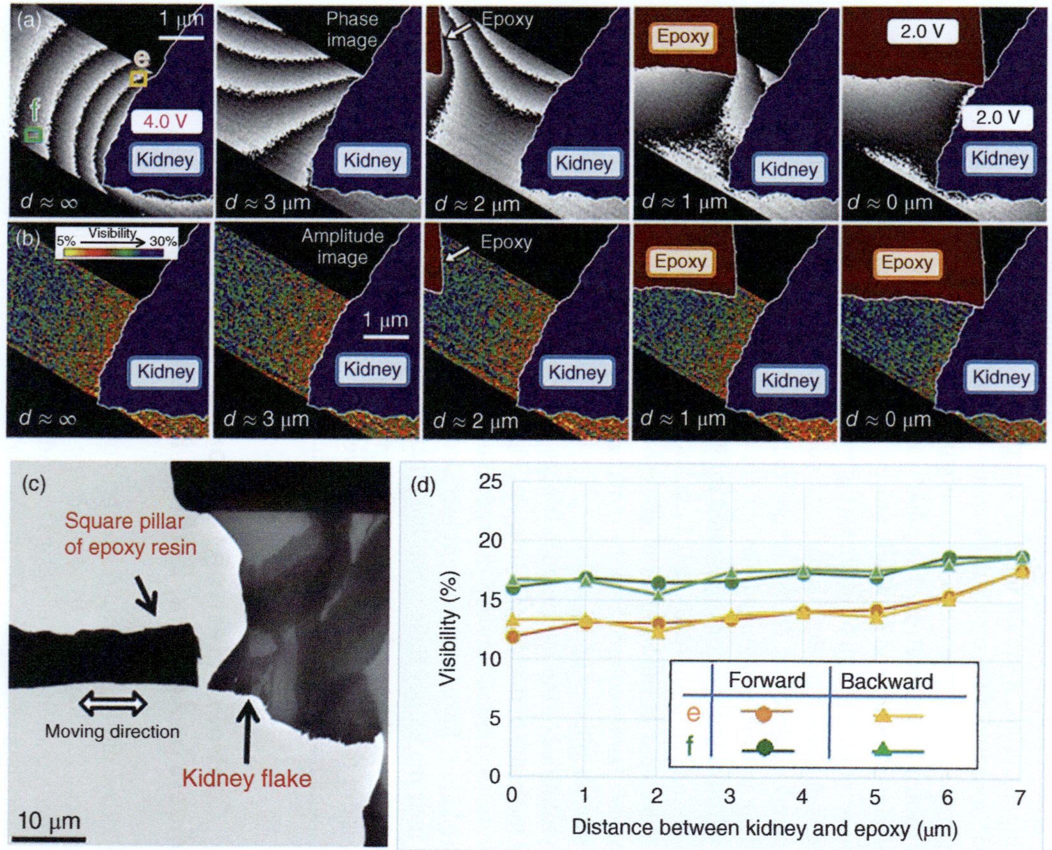

Figure 10.5 (a) Reconstructed phase images when square pillar of epoxy resin was moved to kidney surface. (b) Reconstructed amplitude images when square pillar of epoxy resin was moved to kidney surface. (c) Geometrical configuration of square pillar of epoxy resin and kidney flake. (d) Change in visibility with change in distance between kidney flake and epoxy resin when the pillar was moved forward and backward.

paper by combining CNF and inorganic nanomaterials such as metal nanowires [14–16]. In this section, the collective motions of secondary electrons around CNF were observed by electron holography when a conductive probe was located near the CNF and the bias voltage of the probe was varied [17].

A TEM specimen was prepared by dropping CNF dispersed in ethanol on a microgrid cut in half. The microgrid and a conductive Pt–Ir probe were set on a double-probe piezodriving holder (see Section 6.6). The conductive probe could approach the CNF because the microgrid had been cut in half.

The left panel in Figure 10.6 shows a hologram indicating the location of the CNF and probe in the column of a transmission electron microscope. When holograms were recorded, the area illuminated by the incident electron beam was limited to the CNF and the probe, not to the support film, by using a small condenser aperture to charge the CNF more strongly. If the support film had been illuminated as well, many secondary electrons emitted from the film would have been attracted by the charged CNF, and strong charging would not have been observed. These phenomena were also observed for SiO_2 particles [18].

The right panel in Figure 10.6 shows reconstructed amplitude images when the bias voltage of the probe was increased from −15 to +15 V. The color bar corresponds to the visibility of the fringe contrast in the electron hologram. In this series of reconstructed amplitude images, local fluctuation of the electric field around the CNF is evident when the bias voltage of the probe was −10 and −5 V. We attribute this to the secondary electrons collectively moving around the CNF at these bias voltages. When the bias voltage was less than −10 V, the repelling force on the secondary electrons from the probe would have been too strong to cause collective motions of the secondary electrons around the CNF.

When the bias voltage of the probe was positive, the electric potential difference between the CNF and probe was small. As a result, secondary electrons did not accumulate around the CNF. This means that the bias voltage of the probe affected the motions of the secondary electrons. For microfibrils of sciatic nerve tissue with Os on the surface, electrons tend to move collectively along the specimen surfaces. For CNF, electrons tend to simply accumulate above the specimen surfaces and between the CNF surface and the conductive probe at negative bias voltages. It is expected that cellulose nanopapers consisting of CNFs and metal nanowires, which show

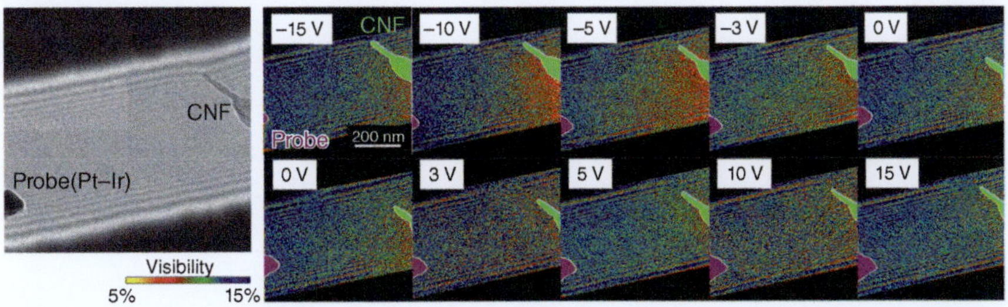

Figure 10.6 (Left) Hologram indicating location of CNF and probe. (Right) Reconstructed amplitude images obtained when bias voltage of probe was increased from −15 V to +15 V. Source: Shindo and Akase [7], with permission from ELSEVIER.

interesting electrical properties such as conductivity [15] and a high dielectric constant [14], will be studied extensively by using in situ electron holography and an applied bias voltage.

References

1 Zhang, X., Hashimoto, T., and Joy, D.C. (1992). *Appl. Phys. Lett.* 60: 784.
2 Yamamoto, T., Hirayama, T., Fukunaga, K., and Ikuhara, Y. (2004). *Nanotechnology* 15: 1324.
3 Szwarcman, D., Lubk, A., Linck, M. et al. (2012). *Phys. Rev. B: Condens. Matter* 85: 134112.
4 Polking, M.J., Han, M.-G., Yourdkhani, A. et al. (2012). *Nat. Mater.* 11: 700.
5 Kuramae, R., Ono, H., Fujikawa, Y. et al. (2012). *Mater. Trans.* 53: 696.
6 Kawamoto, N., Ono, H., Terasaki, Y. et al. (2019). *Mater. Trans.* 60: 2109.
7 Shindo, D. and Akase, Z. (2020). *Mater. Sci. Eng. R Rep.* 142: 100564. https://doi.org/10.1016/j.mser.2020.100564.
8 Akase, Z., Shinmeimae, K., Miyabayashi, Y., et al. (2020). *The 76th Annual Meeting of the Japanese Society of Microscopy*, Osaka (May 25–27).
9 Sato, T., Tsukida, N., Higo, M. et al. (2019). *Mater. Trans.* 60: 2114.
10 Akase, Z., Higo, M., Shimada, K. et al. (2021). *Mater. Trans.* 62: 1589.
11 Fukuzumi, H., Saito, T., Okita, Y., and Isogai, A. (2010). *Polym. Degrad. Stab.* 95: 1502.
12 Yagyu, H., Saito, T., Isogai, A. et al. (2015). *Mater. Interfaces* 7: 22012.
13 Nogi, M., Karakawa, M., Komoda, N. et al. (2015). *Sci. Rep.* 5: 17254.
14 Inui, T., Koga, H., Nogi, M. et al. (2015). *Adv. Mater.* 27: 1112.
15 Koga, H., Nogi, M., Komoda, N. et al. (2014). *NPG Asia Mater.* 6: e93.
16 Celano, U., Nagashima, K., Koga, H. et al. (2016). *NPG Asia Mater.* 8: e310.
17 Akase, Z., Hongo, M., Sato, T., et al. (2019). *62nd Symp. Jpn. Soc. Microscopy*
18 Suzuki, H., Akase, Z., Niitsu, K. et al. (2017). *Microscopy* 66: 167.

11

Extension of Analysis of Collective Motions of Electrons

To extend our analysis of the collective motions of electrons, we first present the observation of electron spin polarization by applying an external magnetic field to electrons accumulated around a charged mica insulator in Section 11.1. In Section 11.2, on the basis of electron holography studies on the accumulation of electrons, which are attracted to positively charged specimens, experiments are designed with an FIB system to observe a similar phenomenon, namely, the dependency of the distribution of secondary electrons around bulk insulator surfaces on the existence of metal elements. In Chapter 12, we discuss the theoretical significance of observing electric charge distributions by electron holography.

11.1 Electron Spin Polarization

So far, we have discussed visualizing the collective motions of electrons around various insulators by detecting the electric field variations due to the electron motions. Although an electron has spin angular momentum $h/(4\pi)$, the direction of the momentum is random for each electron. On the other hand, like the magnetization process of the various magnetic materials discussed in Section 8.2, the spin polarization of electrons accumulated around a charged insulator such as a mica one can be observed by applying an external magnetic field to the electrons as follows.

Suppression of secondary-electron irradiation from a substrate was achieved by fabricating a sharp needle made of a mica specimen, as illustrated in Figure 11.1a, by using an FIB system. No appreciable changes in the distribution of secondary electrons were observed in the reconstructed amplitude images of the mica specimen under external magnetic fields of less than 20 mT. Under these conditions, the electric interactions between the surface of the positively charged mica needle and the surrounding electrons mainly determine their distribution. Unlike the experiments using a gradient in the magnetic field, such as the well-known Stern and Gerlach experiment with Ag atoms [1], a constant magnetic field across the mica needle produced spin polarization torque of electrons without a translational shift.

Furthermore, the magnetic interaction energy $|U_m|$ ($\propto 1/r^3$) between spins with the dipole interaction depending on the distance (r) in the range 0.1–10 nm is much weaker than the electric interaction energy U_e ($\propto 1/r$); i.e. U_e ($=2.31 \times 10^{-19}$ J) is

Material Characterization using Electron Holography, First Edition. Daisuke Shindo and Takeshi Tomita.

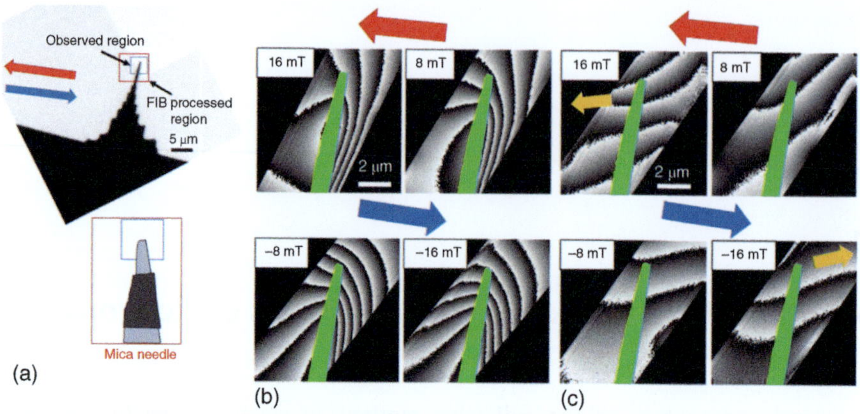

Figure 11.1 (a) Geometric configuration of mica needle fabricated using FIB method. (b) Reconstructed phase images observed under external magnetic fields for which directions are indicated by red and blue arrows. Magnetic flux density of external magnetic field is indicated at top left corner in each image. (c) Magnetic phase information of polarized secondary electrons extracted from reconstructed phase images. Yellow arrows correspond to directions of electron spin polarization observed in reconstructed phase images.

about 10^7 times larger than $|U_\mathrm{m}|$ ($=1.72\times10^{-26}$ J) for $r = 1$ nm [2]. Under this condition, electric field information can easily be separated from magnetic field information: the electric field information obtained without the external magnetic field can be subtracted from the phase information under the external magnetic field condition. The images in Figure 11.1b were obtained when the magnetic fields were applied along the directions indicated by the red and blue arrows. Moreover, the images in Figure 11.1c were obtained by subtracting the electric field information obtained without the external magnetic field from the reconstructed phase image in Figure 11.1b, resulting in revelation of the magnetic phase information due to electron spin polarization under the external magnetic field condition.

Note that the phase image obtained in this manner can provide the difference in magnetic field information due to spin polarization between the region near the mica needle (object-wave region) and that far from the mica needle (reference-wave region). The figure in 11.1c shows that the magnetic information tends to increase with the external magnetic field. It is noted that the observed magnetic information due to electron spin polarization is closely related to the distribution of secondary electrons and to the magnitude of the interaction between the electrons and the charged specimen surface [3, 4].

11.2 Accumulation of Electrons on Bulk Insulator Surface

As noted, the accumulation and collective motions of electrons around various positively charged insulators have been observed experimentally through the amplitude reconstruction process in electron holography. Those experiments were performed by transmission electron microscopy (TEM) using specimens with a thin insulating surface. In this section, we show that similar effects of secondary electrons, which are attracted even to bulk surfaces of insulators, can be observed by scanning ion microscopy (SIM) with an FIB system [5]. The common feature of the phenomena observed by TEM and SIM is the positive charging effect of insulators. During SIM observation, a Ga^+-ion probe is used to scan the bulk insulator surface. At each point in the scanned area, secondary electrons are emitted from the bulk surface while Ga^+ ions accumulate in the bulk specimen, resulting in positive charging (Figure 11.2).

An FIB system was used to deposit two W "islands" on a mica substrate in the form of the character "e" and a superscript minus symbol ("–"). SIM observation was then performed for 48 seconds with a nominal Ga^+ dose of 0.027 ($C\,m^{-2}$), or 0.17 ($ion\,nm^{-2}$). Figure 11.3a shows the initial state. When the "–" island was

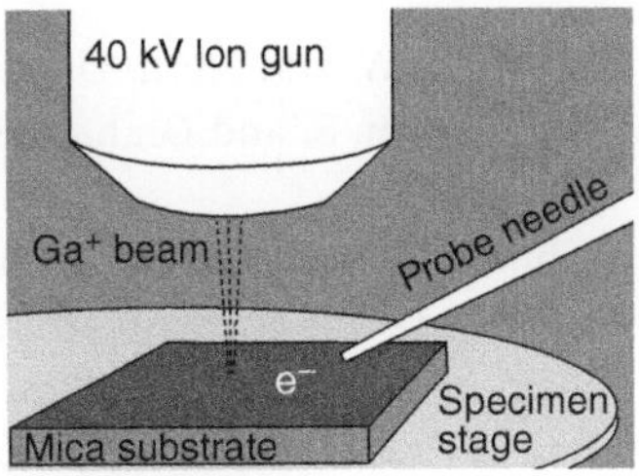

Figure 11.2 Experimental setup in which "e⁻" pattern was drawn on insulating mica surface using W deposition and probe needle.

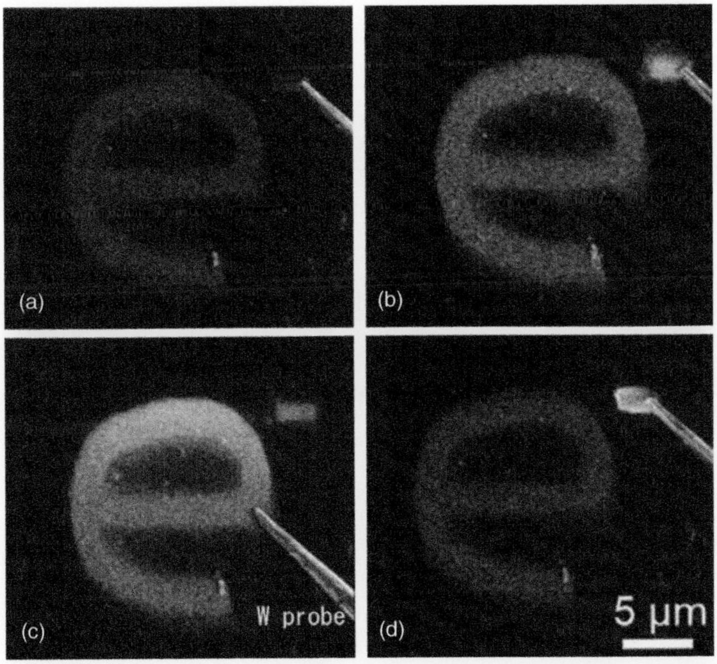

Figure 11.3 Successively recorded SIM images of "e⁻" characters prepared on a mica substrate by W deposition. (a) W islands not touched by +5-V biased probe needle. (b) "–" character touched by the probe. (c) "e" character touched by the probe. (d) "–" character touched by the probe with a 0.17 ion nm^{-2} additional Ga^+ dose after scene (b).

touched with a 5-V biased probe, it became brighter, as did the "e" island (b), indicating charge transfer from the "–" island to the "e" island along with the existence of Ga and W elements on the mica substrate. This matches the results for the epoxy thin film (Figure 10.3) and kidney flake (Figure 10.4). The existence of W was due to its spread during the deposition process. Then, when the "e" island was touched by the probe, it became much brighter, and the "–" island became slightly brighter (Figure 11.3c). Finally, when the "–" island was again touched by the probe, it became brighter, but the "e" island did not (d). The different result shown in (d) compared with those in (b) and (c) is attributed to enhanced sputtering of the specimen surface by Ga^+ ions [5].

References

1 Gerlach, W. and Stern, O. (1922). *Zeitschrift für Phys.* 9: 349.

2 Chikazumi, S. and Graham, C.D. (1997). *Physics of Ferromagnetism*, 2e. Clarendon Press.

3 Shindo, D. and Akase, Z. (2020). *Mater. Sci. Eng., R* 142: 100564.

4 Shindo D. and Sato T.Method for creating electron – beam hologram, magnetic field information measurement method, and magnetic field information measuring device. US Patent 11,067,649, B1 (20, July, 2021)

5 Akase, Z., Higo, M., Shimada, K. et al. (2021). *Mater. Trans.* 62: 1589. https://doi.org/10.2320/matertrans.MT-M2021086.

12

Theoretical Consideration on Visualizing Collective Motions of Electrons

In Chapters 9 and 10, electric charge distributions of secondary electrons around the charged insulators were visualized at nanometer scale by the interference effect of incident electron wave with coherence lengths over micrometer scale. In this section, theoretical basis of observation and evaluation methods of these results is presented. On the theoretical basis, the dual wave–particle features of electrons, i.e. incident plane wave and point charge distribution of secondary electrons, are extensively discussed in the framework of the general theory of relativity.

In Section 12.1, in relation to the wave behavior of electrons, de Broglie's matter wave and wave function in quantum mechanics are outlined. On the basis of quantum mechanics, the electron holographic visualization (disturbance-free observation) technique is explained in Section 12.2. In relation to the coherence lengths depending on the de Broglie wavelength, the electron interference effect is discussed with the Einstein's field equation in the framework of general theory of relativity in Section 12.3. Based on these theoretical considerations, a model (spinning linear wave) for the electron interference effect is presented in Section 12.4. With this model for the electron interference effect, intensity distributions of electron diffraction and interference fringes are formulated, being consistent with the formulations with the plane wave incidence for sufficiently large number of incident electrons in Section 12.5. In Section 12.6, the wave–particle dualism is interpreted with the proposed model.

12.1 De Broglie's Matter Wave and Wave Function

The wave nature of the electron was first presented by de Broglie in 1923 [1]. He considered an electron with a stationary harmonic oscillation represented by $\sin(2\pi\nu t')$, where t' is the time and ν is the frequency proposed by de Broglie, with the relation [1, 2]

$$mc^2 = h\nu \tag{12.1}$$

where m is the rest mass of the electron. By applying the Lorentz transformation to the harmonic oscillation at a particle position for an observer in an inertia system

Material Characterization using Electron Holography, First Edition. Daisuke Shindo and Takeshi Tomita.

moving in direction z with speed v, de Broglie derived [1]

$$\sin(2\pi\nu t') = \sin\left[\frac{2\pi\nu}{\sqrt{1-\frac{v^2}{c^2}}}\left(t-\frac{vz}{c^2}\right)\right]$$
$$= \sin(\omega t - kz)$$
$$\left(\omega = \frac{2\pi\nu}{\sqrt{1-\frac{v^2}{c^2}}}, k = 2\pi/\lambda\right) \tag{12.2}$$

where c is the speed of light and λ (= h/p, h: Planck's constant) is the de Broglie wavelength. The right-hand side of the last line of Eq. (12.2) represents the matter wave propagating along the z-direction.

In *The Evolution of Physics*, Einstein and Infeld referred to de Broglie's matter wave as follows [3]:

> ...The mass, charge, and velocity of each individual electron is known. If we wish to associate in some way a wave concept with a uniformly moving electron or electrons, our next question must be: what is the wave-length? This is a quantitative question and a more or less quantitative theory must be built up to answer it. This is indeed a simple matter. The mathematical simplicity of de Broglie's work, providing an answer to this question, is most astonishing. At the time his work was done, the mathematical technique of other physical theories was very subtle and complicated, comparatively speaking. The mathematics dealing with the problem of waves of matter is extremely simple and elementary but the fundamental ideas are deep and far-reaching....
> from "The Evolution of Physics" by A. Einstein and L. Infeld, Cambridge University Press, Cambridge, 2nd ed. (1978) p. 277.

On the basis of de Broglie's matter wave, a plane wave of incident electrons propagating along the z-direction is represented in quantum mechanics as a wave function in complex notation:

$$\psi_{\mathrm{PW}}(\vec{z}, t) = \exp[i(\vec{k}\cdot\vec{z} - \omega t)] \tag{12.3}$$

The wave function of this plane wave satisfies the time-independent Schrödinger equation and is utilized extensively to analyze various electron interference effects (see Chapter 3). A characteristic feature of electron holographic visualization, disturbance-free observation, is explained in the next section.

12.2 Disturbance-Free Observation

Taking into account the experimental results of observing electron motions by using simulation (Chapters 9 and 10), here we discuss the theoretical background for visualization of electron motions by utilizing the interference effect of incident electrons in the framework of quantum mechanics [4].

To date, the orbits or trajectories of electrons and other particles have been macroscopically investigated by indirect methods utilizing inelastic scattering processes. However, such processes result in decoherence [5], meaning that the measurement apparatus disturbs the system to be observed. In electron holography, the measurement apparatus is elastic electrons, and the object to be observed is the electric field of the system, which consists of a charged specimen and secondary electrons with various kinds of interactions, including inelastic scattering with incident electrons.

Figure 12.1 shows the experimental setup used for holographic visualization of electron motions. In the analysis of electron holographic visualization, we can assume a stationary state in which the charged specimen and the motions of secondary electrons around it are static because of the high velocity of the incident electrons, which are accelerated at a few hundred kV. Therefore, the interaction of the incident electrons with a system consisting of a charged specimen and secondary electrons is generally given by a time-independent Schrödinger equation

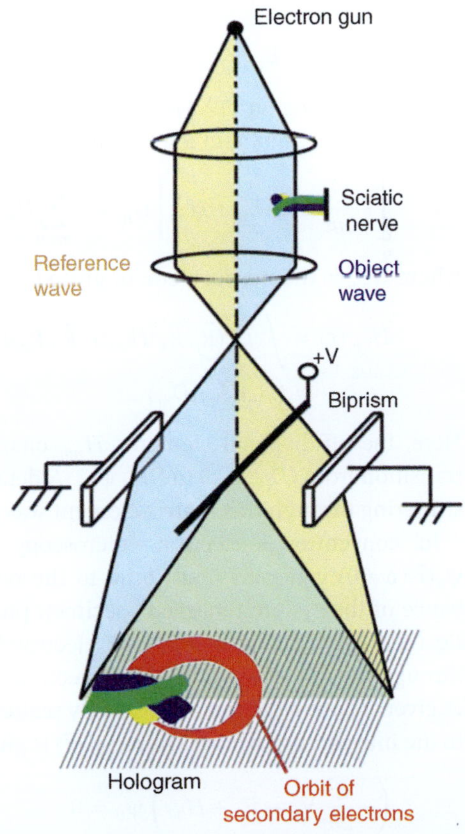

Figure 12.1 Schematic illustration of experimental setup used to form hologram of microfibrils of sciatic nerve tissue.

with the Hamiltonian H as

$$\left(-\frac{\hbar^2}{2m}\nabla^2 + H_s + H_{\text{int}}\right)\Psi = E\Psi \tag{12.4}$$

where $\hbar$ is Planck's constant divided by 2π and m is electron mass. The three terms in parentheses correspond, respectively, to the kinetic energy of the incident electrons, the energy of the system (the charged specimen plus secondary electrons), and the interaction energy between the incident electrons and the system. The total wave function for the incident electrons and the system can be expanded in a series:

$$\Psi(\vec{r}, \vec{R}_j, \vec{R}_k) = \sum_n \psi_n(\vec{r}) D_n(\vec{R}_j, \vec{R}_k) \tag{12.5}$$

where $\psi_n(\vec{r})$ is the wave function of the incident electrons; $\psi_0(\vec{r})$ and $\psi_{n\neq 0}(\vec{r})$ correspond to its elastic scattering part and inelastic scattering part, respectively; D_n is the wave function of the system representing the nth excited state with energy ε_n; and $\vec{R}_j$ and $\vec{R}_k$ represent the positions of constituent charges (electrons (j) and nuclei (k)). D_n is determined by

$$H_s D_n = \varepsilon_n D_n \tag{12.6}$$

The energy loss of the inelastically scattered electrons with wave function $\psi_n(\vec{r})$ is given by

$$\Delta E = E - E_n = \varepsilon_n \tag{12.7}$$

By substituting Eq. (12.5) into Eq. (12.4), multiplying with D_n^*, and integrating over $\vec{R}_j$ and $\vec{R}_k$, we can obtain a set of equations for $\hbar$:

$$\left[-\frac{\hbar^2}{2m}\nabla^2 - E_n + H_{nn}\right]\psi_n = -\sum_{m\neq n} H_{nm}\psi_m, n = 0, 1, \ldots \tag{12.8}$$

where H_{nm} is the matrix element given by

$$\begin{aligned} H_{nm}(\vec{r}) &= \int D_n^*(\vec{R}_j, \vec{R}_k) H_{\text{int}}(\vec{r}, \vec{R}_j, \vec{R}_k) D_m(\vec{R}_j, \vec{R}_k) d^3R_j d^3R_k \\ &= \langle D_n | H_{\text{int}} | D_m \rangle \end{aligned} \tag{12.9}$$

Here, the off-diagonal elements H_{nm} characterize the probability of an inelastic transition from (D_m, ψ_m) to (D_n, ψ_n). A detailed discussion on the effect of inelastic scattering through this matrix element was given by Yoshioka [6].

In conventional electron microscopy, the inelastically scattered electrons $\psi_n(\vec{r})(\text{n} \neq 0)$ generally contribute to the observed intensity, resulting in a disturbance of the system (charged specimen plus secondary electrons) as expressed by Eq. (12.9). In electron holography, electron holograms are obtained only from $\psi_0(\vec{r})$ through the interference effect between the objective wave and the reference wave, as given below. Note that inelastically scattered wave $\psi_n(\vec{r})(n \neq 0)$ cannot contribute to the interference fringes. Here, $\psi_0(\vec{r})$ is given by simplifying Eq. (12.8):

$$\left(-\frac{\hbar^2}{2m}\nabla^2 - E_0 + H_{00}\right)\psi_0 = 0 \tag{12.10}$$

where H_{00} is potential energy $-e\varphi$ (φ is the scalar or electric potential). As solutions of Eq. (12.10), two components of $\psi_0(\vec{r})$ in electron holography are the object wave and the reference wave. Quantum non-demolition measurements [7–9] and negative-result measurements [10] for photons and spins have been extensively discussed as disturbance-free measurements in the field of quantum measurements. In electron holography, the motions of electron-induced secondary electrons can be located through electric field visualization without disturbing their motions since electromagnetic fields are completely independent of charges and currents. It is also noted that the current of the secondary electrons in stationary motions observed by electron holography is considerably larger than that of the incident electrons.

12.3 Electron Interference and General Relativity

On the basis of experiments carried out by the disturbance-free observation, two characteristic features of electrons in such experiments (Figure 12.1) have been pointed out [11]. The first is related to secondary electrons (orbit indicated in red), for which nanometer-scale motion is visualized by electron holographic disturbance-free observation. The second is related to incident electrons (indicated by blue and yellow shading) for which observation is done using interference effects extending to the micrometer scale. We classify and briefly discuss the two characteristic features of electrons:

(1) Particle-like feature: electromagnetic field formation of secondary electrons visualized as a point charge at nanometer-scale resolution.
(2) Wave-like feature: interference effects of incident electrons with coherence lengths over the micrometer scale.

Note that the interference effect of the electrons depends on their momentum while their charge does not depend on their momentum.

The orbits of secondary electrons revealed at the nanometer scale by electron holographic visualization are well interpreted by Maxwell's equations with a point-charge picture of the electron through comparison between simulated and experimental results, as described in Section 9.1. The experimental results are consistent with the charge conservation law at the nanometer scale. In other words, the electric field produced by electric charges observed by electron holography corresponds well to Maxwell's equations and is consistent with the special theory of relativity [12].

In contrast, incident electrons with the interference effect show wave-like features over the micrometer scale. The range of interference regions or coherent regions is characterized by the coherence lengths, as given in Section 5.3.

Note that the coherence lengths depend on the momentum and are basically applicable to any particle, including photons and neutrons, regardless of whether the particle has a charge. Thus, unlike the electric field treated by Maxwell's equations, it was noticed that the interference effect cannot be discussed in the framework of the special theory of relativity [11]. Here, we recall the words of Einstein and Infeld given in Chapter 1:

"...it is not the charges nor the particles but the field in the space between the charges and particles which is essential...." Taking this into account, we have to discuss the field corresponding to the wave features of the electron depending on the mass or momentum through Einstein's equations in the framework of the general theory of relativity [13–15]. These interesting features of the electron, noticed through electron holographic visualization of collective electron motions, are discussed theoretically in the following sections.

12.3.1 Einstein's Field Equations Based on General Relativity

To discuss the gravitational field around an electron on the basis of Einstein's field equations in the framework of the general theory of relativity, we first outline Einstein's field equations and discuss some of their characteristic features in comparison with the special theory of relativity.

As discussed in Section 2.2, Maxwell's equations for an electromagnetic field have been reformulated on the basis of the special theory of relativity. However, if there is gravity, such fields cannot be discussed in the special theory of relativity that is the case for inertia systems without acceleration. Electromagnetic fields in a gravitational and/or acceleration system should be discussed in the general theory of relativity with a four-dimensional Riemannian space.

The squared magnitude of line element ds given by the Minkowskian metric tensor $\eta_{\mu\nu}$ defined in Eq. (2.16) for the Minkowskian space can be generalized to the metric tensor $g_{\mu\nu}$ for the Riemannian space given by

$$(ds)^2 = g_{\mu\nu} dx^\mu \, dx^\nu \tag{12.11}$$

where $g_{\mu\nu}$ generally changes with the four-dimensional coordinates. The "principle of equivalence" means that, for any Riemannian space characterized by Eq. (12.11), we can always set a tiny Minkowskian space by controlling acceleration.

Einstein's field equations are given by [14, 15]

$$R_{\mu\nu} - \frac{1}{2} R g_{\mu\nu} = -\kappa T_{\mu\nu}$$
$$(\mu, \nu = 0, 1, 2, 3) \tag{12.12}$$

where $R_{\mu\nu}$ is the contracted curvature tensor, $T_{\mu\nu}$ is the energy–momentum tensor, κ is Einstein's constant of gravitation, and R is the curvature scalar ($R = g^{\mu\nu} R_{\mu\nu}$). These equations are nonlinear partial differential equations in terms of $g_{\mu\nu}$, so they are generally difficult to solve.

The special theory of relativity is considered to be a special case of the general theory of relativity in which $g_{\mu\nu}$ have constant values, given as Eq. (2.16). This condition corresponds to neglecting the effect of gravity. By considering the case in which $g_{\mu\nu}$ differ from constant values by small magnitudes compared with unity, we can derive Newton's theory of gravitation [13].

After presenting his special theory of relativity in 1905 [16] and the general theory of relativity in 1915 [13, 17], Einstein discussed the stability of a charged particle (electron) in 1919 [18]. He modified his gravitational equation (Eq. (12.12)) to take into account the charged particle by including Maxwell's energy–momentum tensor (see Eq. (2.28) in Section 2.2) on the right side of Eq. (12.12). By assuming that the stability of the particle is solely due to gravity and that Maxwell's energy–momentum tensor is proportional to some quantity consisting of $g_{\mu\nu}$ (metric tensor) and its derivatives in terms of four-dimensional time–space coordinates x^μ ($x^0 = ct, x^1 = x, x^2 = y, x^3 = z$), Einstein derived the relation

$$f_{\mu\nu} j^\nu + \frac{1}{4\kappa}\frac{\partial R}{\partial x^\mu} = 0 \tag{12.13}$$

where $f_{\mu\nu}$ is an antisymmetric covariant tensor of the second rank corresponding to the electric and magnetic field components, j^ν is a contravariant four-vector current, R is a curvature scalar, and κ is the constant that is given as $8\pi K/c^2$ (K: universal gravitational constant). The equation shows that the Coulomb repulsive force (the first term) is offset by the gravitational compressive force (the second term). In other words, it demonstrates that the gravitational compressive force and the Coulomb repulsive force are balanced.

The shape of a linear wave is closely related to the gravitational potential distribution obtained by Infeld and Schild, as shown in the next section, as one solution to Einstein's field equation for neutral matter. This theoretical indication that the stabilization of the electron is solely due to gravity thus agrees well with the fundamental feature of the coherence length, which is related to gravity and not the charge, as discussed in the next section.

In "On the Generalized Theory of Gravitation," Einstein emphasized the importance of the general theory of relativity [19]:

> I do not see any reason to assume that the heuristic significance of the principle of general relativity is restricted to gravitation and that the rest of physics can be dealt with separately on the basis of special relativity, with the hope that later on the whole may be fitted consistently into a general relativistic scheme. I do not think that such an attitude, although historically understandable, can be objectively justified. The comparative smallness of what we know today as gravitational effects is not a conclusive reason for ignoring the principle of general relativity in theoretical investigations of a fundamental character. In other words, I do not believe that it is justifiable to ask: What would physics look like without gravitation?
> From "On the Generalized Theory of Gravitation" by A. Einstein, in Scientific American resource library: readings in the physical sciences, vol. 1, W. H. Freeman (1969) p. 64.

12.3.2 Infeld and Schild's Approximate Solution to Einstein's Field Equations

In 1938, Einstein, Infeld, and Hoffmann [20] asserted that the energy–momentum tensor in Eq. (12.12) must be regarded as purely temporary and more or less as a phenomenological entity for representing the structure of matter. They also noted that its entry into the equations made it impossible to determine how much the results obtained were independent of the particular assumption made regarding the constitution of matter. They thus presented a different approach to solving the gravitational field equations – they neglected the energy–momentum tensor and assumed that each particle is a singularity of the gravitational field. They showed that the motions of particle singularities are completely determined by the gravitational field equations for empty space, i.e.

$$R_{\mu\nu} - \frac{1}{2}Rg_{\mu\nu} = 0 \tag{12.14}$$

As mentioned above, the coherence length and wave behavior of an electron depend on its wavelength or momentum and not its charge, so these features should be discussed on the basis of Einstein's field equations.

In 1949, Infeld and Schild [21] used Eq. (12.14) in a study of the motion of test particles and discussed the gravitational field corresponding to N_0 point particles by assuming that the gravitational field dependence on parameter m_0 corresponds to a mass with $m_0 = 1/N_0$; each particle was represented by a singularity of the gravitational field, as shown in Figure 12.2a. In their discussion, they assumed neutral matter. This assumption is applicable to the following discussion of the electron field because the wave field of an electron depends on its mass or momentum and not its charge, as noted above. They expanded $g_{\mu\nu}$ in a series in terms of m_0:

$$g_{\mu\nu} = \eta_{\mu\nu} + a_{\mu\nu} + (m_0)b_{\mu\nu} + (m_0)^2c_{\mu\nu} + \dots \tag{12.15}$$

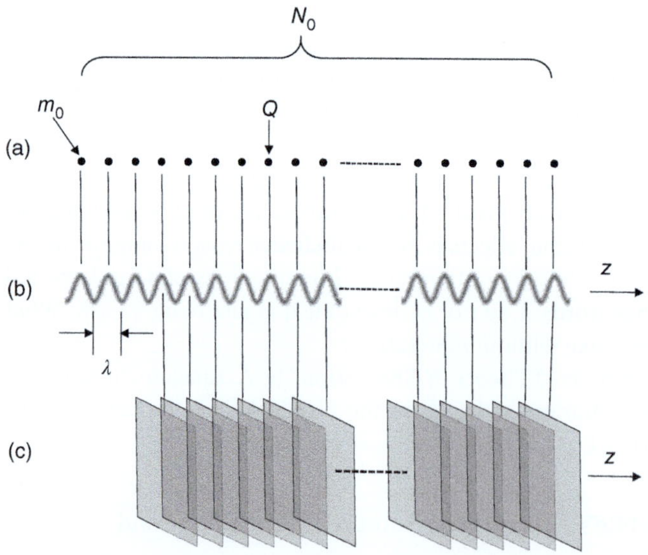

Figure 12.2 (a) Illustration showing electron consisting of one-dimensional distribution of N_0 point particles used in discussion of gravitational field equations. (b) One-dimensional gravitational field distribution with wavelength λ corresponding to N_0 point particles. (c) Schematic illustration showing plane wave consisting of gray planes with same phase.

where $\eta_{\mu\nu}$ is the Minkowskian metric given by Eq. (2.16). They inserted $g_{\mu\nu}$ of Eq. (12.15) into Eq. (12.14). In the limit of $m_0 \to 0$ ($N_0 \to \infty$), the higher-order terms of $(m_0)^n$ ($n \geq 2$) were neglected to obtain the field equations

$$B_{\mu\nu} = F_{\mu\nu} + L_{\mu\nu} = 0 \tag{12.16}$$

in terms of $\beta_{\mu\nu}$, which is introduced in the form of equivalent relations:

$$\beta_{\mu\nu} = b_{\mu\nu} - \frac{1}{2}\eta_{\mu\nu}\eta^{\rho\sigma}b_{\rho\sigma}, b_{\mu\nu} = \beta_{\mu\nu} - \frac{1}{2}\eta_{\mu\nu}\eta^{\rho\sigma}\beta_{\rho\sigma} \tag{12.17}$$

In Eq. (12.16), $F_{\mu\nu}$ involves the second derivatives of $\beta_{\mu\nu}$ while $L_{\mu\nu}$ contains non-linear terms of the products of the derivatives of $\beta_{\mu\nu}$ and $a_{\mu\nu}$. In the limit of $m_0 \to 0$ around point Q (Figure 12.2a), they obtained field equations containing linear terms by neglecting $L_{\mu\nu}$, i.e.

$$F_{\mu\nu} = 0 \tag{12.18}$$

The gravitational field equations of Eq. (12.18) can be finally written as linear equations:

$$\eta^{\rho\sigma}\frac{\partial}{\partial x^\rho}\frac{\partial}{\partial x^\sigma}\beta_{\mu\nu} = 0 \tag{12.19}$$

$$\eta^{\rho\sigma}\frac{\partial}{\partial x^\sigma}\beta_{\mu\rho} = 0 \tag{12.20}$$

$$(\sigma,\ \rho = 0, 1, 2, 3; x^0 = \mathrm{ct}, x^1 = \mathrm{x}, x^2 = \mathrm{y}, x^3 = \mathrm{z})$$

These two equations have structures similar to those of Maxwell's equations for electromagnetic potentials (Eqs. (2.22) and (2.23)) without a charge and current under the condition $m_0 \to 0$ ($N_0 \to \infty$). Therefore, the solution of these equations corresponds to a waveform similar to that of an electromagnetic wave.

12.4 Spinning Linear Wave Model

Taking into account the result obtained by Infeld and Schild discussed in the section above, we consider here the finite conditions for the electron; N_0 is a large number but not infinity, i.e. $N_0 \to N_e$ and thus $m_0 \to m_e$ accordingly with the relation

$$m_e \cdot N_e = 1 \tag{12.21}$$

Under these conditions, we assume that the form of the matter wave given by the last term in Eq. (12.2) is a solution of these equations for the electron; the waveform, with a phase velocity of c^2/v, for the electron in Eq. (12.2) is the same as that of the electromagnetic wave with phase velocity c. In the derivation of Eqs. (12.19) and (12.20), the electron is considered to consist of a one-dimensional distribution of particles (Figure 12.2a), so its gravitational field is also linear along the z-direction. Given the linearity of the gravitational field, we assume that the corresponding wave function is linear and thus given in a one-dimensional form:

$$\psi_{\mathrm{LW}}(z, t) = \exp[i(kz - \omega t)] \tag{12.22}$$

for the incident electron in the following discussion of quantum mechanical treatment. This one-dimensional wave function satisfies the Schrödinger equation as does the wave function in Eq. (12.3). For a fixed t (= 0) of the linear wave ψ_{LW}, $\cos(kz)$ is illustrated as a function of z in Figure 12.2b while the regularly distributed wavefronts of the plane wave ψ_{PW} in Eq. (12.3) are illustrated as blue planes in (c) for comparison. We assume that each linear wave rotates or spins quickly around the particle position (the center of the linear wave) due to the quantized spin angular momentum with a magnitude of $h/(4\pi)$. Therefore, due to this quick spinning motion, the high density of the spinning linear wave forms around the particle position, and thus it exhibits a particle feature when observed on a relatively large time scale.

While the plane wave is infinitely spread and thus has an infinite coherence length, the length of the linear wave is given by

$$L_{\text{LW}} = N_e \cdot \lambda \tag{12.23}$$

Figure 12.3a schematically illustrates the size and motion of a spinning linear wave, where the direction of its incidence parallel to the z-direction is indicated by a thick red arrow, and the spinning motion of the linear wave is represented by thin red arrows.

Since a linear wave is assumed to be produced from an ideal monochromatic point source with no energy spread, as the case of a plane wave, the length of a linear

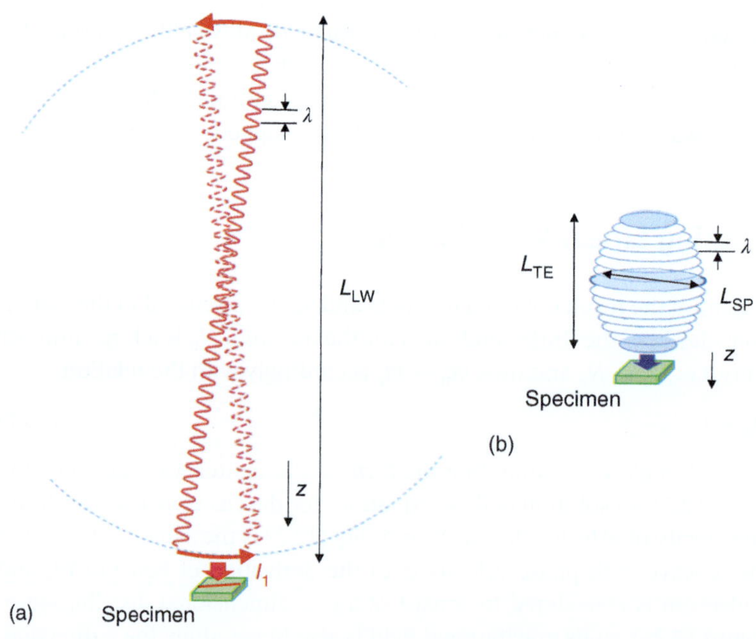

Figure 12.3 (a) Illustration showing motion of linear wave with length L_{LW}. Spinning motion is indicated by thin red arrows; direction of electron incidence parallel to z-direction is indicated by thick red arrow. (b) Illustration of wave packet moving along z-direction. Sizes of spatial and temporal coherence lengths are L_{SP} and L_{TE}, respectively.

wave (L_{LW}) is considered to exceed effective coherence lengths of spatial or lateral coherence length (L_{SP}), temporal or longitudinal coherence length (L_{TE}) as noted in Section 5.3. Since the effective coherence lengths at the micrometer scale are much greater than the length of the interatomic distance (L_{ID}) at the nanometer scale, we have

$$L_{PW}(\infty) > L_{LW} > L_{SP}, L_{TE} \gg L_{ID} \tag{12.24}$$

where the size of the infinitely spreading plane wave is indicated by $L_{PW}(\infty)$. Since the length of the linear wave (L_{LW}) is much greater than the scale of the interatomic distance (L_{ID}), the path traced by the spinning motion of the incident linear wave on the specimen is a straight line (red line l_1 in Figure 12.3a).

In the following sections, the incidences of many electrons with spinning motions in random directions are shown to accumulate and form trace lines that cover the specimen surface and the wire plane. This results in the formulations of diffraction intensities and interference fringes, which are equivalent to those formulated with the plane wave incidences.

12.5 Electron Interference Formulated with Spinning Linear Wave

In the following discussion, we consider the formation of diffraction patterns and interference fringes in accordance with the geometrical configuration of a transmission electron microscope as schematically illustrated in Figure 12.4. The incident electrons can be treated as a plane wave even for a relatively short distance between the electron gun and the specimen due to the condenser lens system. A diffraction pattern can be observed near the specimen position (back focal plane) due to the objective lens system.

12.5.1 Interpretation of Diffraction Intensity

When high-energy electrons are incident on a thin material with thickness Δz in the object plane, as illustrated in Figure 12.4, the specimen can be represented as [22, 23]

$$q(x, y) = \exp[i\sigma\varphi(x, y)\Delta z] \tag{12.25}$$

where the effect of the specimen on the incident plane wave is a phase shift under the phase object approximation. In Eq. (12.25), the integration of the phase shift due to the scalar potential along the z-direction is replaced with the projected two-dimensional scalar potential distribution with specimen thickness Δz, i.e.

$$\sigma \int_0^{\Delta z} \varphi(x, y, z)dz = \sigma \cdot \varphi(x, y) \cdot \Delta z \tag{12.26}$$

where σ is the interaction constant given by

$$\sigma = \frac{2\pi}{\lambda V\left(1 + \sqrt{1 - \frac{v^2}{c^2}}\right)} \tag{12.27}$$

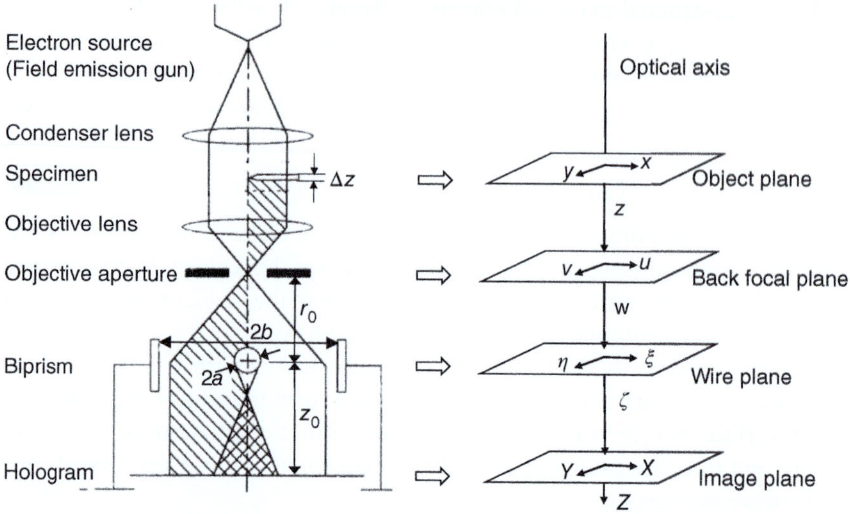

Figure 12.4 Geometrical configuration of transmission electron microscope with corresponding coordinate systems on right. Shaded region indicates object wave.

The V and v are the electron acceleration voltage and electron velocity, respectively. The scattering amplitude is given by

$$\psi(u, v, t) = C \exp(-i\omega t) \iint q(x, y) \exp(-2\pi i(ux + vy))\, dxdy \tag{12.28}$$

where C is a constant and u, v are the coordinates of the back focal plane shown in Figure 12.4. The double integrals correspond to the two-dimensional Fourier transformation [22, 23]. Therefore, the relative diffraction intensity distribution is independent of time t and is thus given sa

$$\psi^*(u, v)\psi(u, v) = \left| \iint q(x, y) \exp(-2\pi i(ux + vy))\, dxdy \right|^2 \tag{12.29}$$

For comparison with the incidence of the plane wave shown in Figure 12.5a, the motion of an incident spinning linear wave is illustrated in (b). In both figures, thick arrows indicate the directions of the incident and diffracted electrons while thin arrows correspond to the translational shift due to the spinning motion of the linear wave in (b).

When the plane wave incidence satisfies the Bragg condition, i.e. $2d\sin\theta = n\lambda$, the interference effect is enhanced, resulting in strong scattering intensity (Figure 12.5a). The phases of the incident plane wave at positions p_1 and p_2 are the same, and those of the diffracted plane wave at positions p_3 and p_4 are also the same. In contrast, when a linear wave moves from the left to right positions, the phases of the linear wave at positions q_1 and q_2 are generally different due to the movement parallel to the thick arrows in the beam direction.

When a linear wave satisfies the Bragg condition, the phases of the diffracted linear wave at positions q_3 and q_4 are also different, but the path difference of the linear wave at the left ($q_1 - q_3$) and right ($q_2 - q_4$) positions is wavelength λ. Therefore, the Bragg condition for the linear wave corresponds to the maintenance of its waveform,

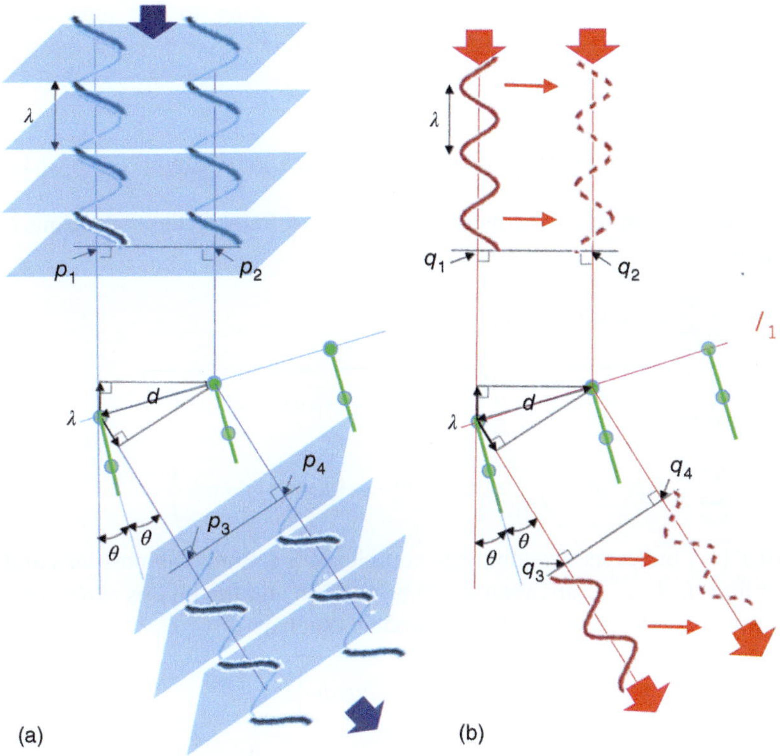

Figure 12.5 (a) Illustration showing geometrical configuration of incident and diffracted plane waves under Bragg condition in object plane. Directions of waves are indicated by thick blue arrows. Green circles correspond to atoms. (b) Illustration showing diffraction process of spinning linear wave in object plane. Directions of incident and diffracted waves are indicated by thick red arrows. Thin red arrows indicate direction of translational motion of linear wave due to its spinning motion.

as shown in Figure 12.5b. The maintenance of the waveform of the linear wave depends on the geometrical conditions of the Bragg diffraction angle θ, the interatomic distance d, and the wavelength λ and not on the phase change due to movement along the incident beam direction.

The maintenance of the waveform of the linear wave is independent of the phases at positions p_1 and p_2, so the degree of waveform maintenance of the linear wave along the line trace (l_1 in Figure 12.3a) can be simply evaluated as the magnitude of the interference effect when the phases of the linear wave along this line are assumed to be equal. For an incident spinning linear wave, the Bragg diffraction intensity is thus evaluated using Eq. (12.29), assuming the area for the two-dimensional integral to be a narrow band region, i.e. l_1 in Figures 12.3a and 12.5b. Therefore, the scattering amplitude of a spinning linear wave corresponding to the first incident electron is given similarly to Eq. (12.28):

$$\psi_1(u, v, t) = C\exp[-i\omega t_1]\iint_{l_1} q(x, y)\exp(-2\pi i(ux + vy))dxdy \tag{12.30}$$

where $\iint_{l_1} \ldots$ is the domain of integration corresponding to the area of l_1 (Figures 12.3a and 12.5b). The scattering amplitude for an electron of the hth incidence $\psi_h(u, v, t)$ is thus generally given as

$$\psi_h(u, v, t) = C\, exp[-i\omega t_h] \iint_{l_h} q(x, y) \exp(-2\pi i(ux + vy))\, dxdy \tag{12.31}$$

The relative diffraction intensity distribution averaged for n incident electrons is finally given by

$$\begin{aligned} I(u, v) &= \frac{1}{n}\sum_{h=1}^{n} \psi_h^*(u, v, t) \cdot \sum_{k=1}^{n} \psi_k(u, v, t) \\ &= \frac{1}{n}\sum_{h=1}^{n} \overline{\left\{ \exp(-i\omega t_h) \iint_{l_h} q(x, y) \exp[-2\pi i(ux + vy)] dxdy \right\}} \cdot \\ &\quad \sum_{k=1}^{n} \left\{ \exp(-i\omega t_k) \iint_{l_k} q(x, y) \exp[-2\pi i(ux + vy)] dxdy \right\} \end{aligned} \tag{12.32}$$

The overbar $\overline{\{\ldots\}}$ indicates the conjugate complex value. Since the positions and directions of lines $l_1, l_2, \ldots l_\mathrm{n}$ are assumed to be random and the phases of $\omega t_1, \omega t_2, \ldots \omega t_\mathrm{n}$ are uncorrelated, the summation of products such as

$$\begin{aligned} &\overline{\left\{ \exp(-i\omega t_h) \iint_{l_h} q(x, y) \exp[-2\pi i(ux + vy)] dxdy \right\}} \\ &\quad \cdot \left\{ \exp(-i\omega t_k) \iint_{l_k} q(x, y) \exp[-2\pi i(ux + vy)]\, dxdy \right\} \end{aligned} \tag{12.33}$$

tends to vanish, except the summation for $h = k$. Therefore, we obtain

$$\begin{aligned} I(u, v) &= \frac{1}{n}\sum_{h=1}^{n} \overline{\left\{ \iint_{l_h} q(x, y) \exp[-2\pi i(ux + vy)] dxdy \right\}} \cdot \\ &\quad \left\{ \iint_{l_h} q(x, y) \exp[-2\pi i(ux + vy)] dxdy \right\} \\ &\approx \left| \iint_{S} q(x, y) \exp[-2\pi i(ux + vy)] dxdy \right|^2 \end{aligned} \tag{12.34}$$

The last line of this equation can be obtained if we consider $\sum_{h=1}^{n} \iint_{l_h} \ldots$ to be equal to $\iint_S \ldots$ for a large number n, where $\iint_S \ldots$ is the domain of integration corresponding to the area of the two-dimensional specimen surface. Therefore, the equation shows that the formulation of the intensity distribution for electron diffraction using spinning linear waves is equivalent to the formulation in Eq. (12.29), obtained with simple plane wave incidence when the number of incident electrons is sufficiently large.

12.5.2 Interpretation of Interference Fringes

Here we discuss the formation of interference fringes with spinning linear waves. The two-dimensional incident wave around a wire plane is schematically shown in

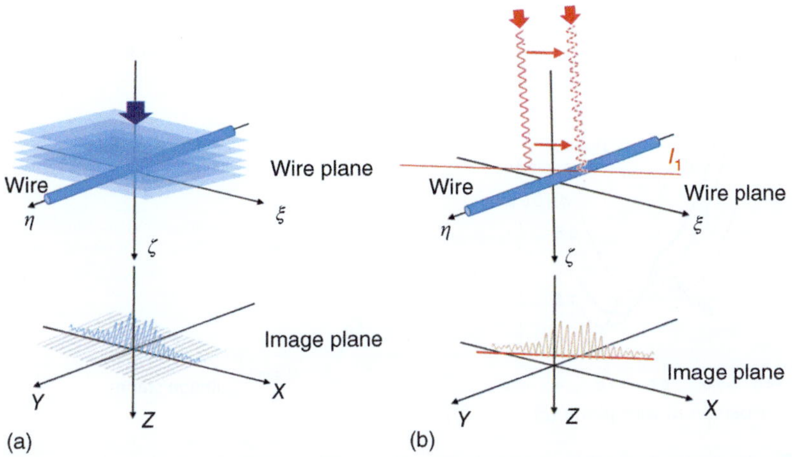

Figure 12.6 Illustration showing geometrical configuration of incident electron beam and biprism wire. (a) Two-dimensional wave incidence. Formation of interference fringes is illustrated at bottom. (b) Spinning linear wave crosses wire plane perpendicularly. Expected intensity profile of part of the interference fringes is shown at bottom.

the upper part of Figure 12.6a. The formation of interference fringes in the image plane is illustrated in the lower part. We assume that the motion of a spinning linear wave is the translational shift indicated by the thin red arrows crossing the ξ-η plane in (b). The expected intensity profile of a part of the interference fringes for the spinning linear wave is illustrated in the lower part of (b). In accordance with the treatment by Komrska [24] outlined in Section 5.4.2, the amplitude of the interference fringes for the two-dimensional incident wave is given by

$$\psi(X,Y,t) = D\exp(-i\omega t)\int_{-\eta_{max}}^{\eta_{max}}\left[\int_{-\xi_{max}}^{a} + \int_{a}^{\xi_{max}}\right]\exp\{i[\phi(\xi,\eta)+\phi_U(\xi,\eta)]\}d\xi d\eta \tag{12.35}$$

where D is a constant and the phase factor caused by the path difference $[\phi\ (\xi,\eta)]$ and the electric field around the wire $[\phi\ _U(\xi,\eta)]$ is given as

$$\begin{aligned}
&\phi(\xi,n)+\phi_U(\xi,n)\\
&\quad = k_0\left\{\frac{1}{2}\left(\frac{1}{r_0}+\frac{1}{z_0}\right)\left[\left(\xi-\frac{Xr_0}{r_0+z_0}\right)^2+\left(\eta-\frac{Yr_0}{r_0+z_0}\right)^2\right]+\frac{|e|\varphi_W}{\mathrm{In}(a/b)}\frac{1}{2E}\right\}\\
&\qquad \cdot\xi\left[\left(\frac{\xi}{r_0}+\frac{\xi-x}{z_0}\right)\mathrm{In}\frac{|\xi|}{b}+\pi\ \mathrm{sgn}\ \xi-\frac{\xi}{b}\left(1+\frac{1}{12}\frac{\xi^2}{b^2}\right)\right]
\end{aligned} \tag{12.36}$$

where E is the energy of the incident electrons and the other parameters (a, b, r_0, etc.) are as shown in Figure 12.4 and listed in Table 5.1. The integration in Eq. (12.35) is carried out using coordinates ξ, η over the entire area, excluding the wire region. The relative intensity distribution of the interference fringes with the two-dimensional

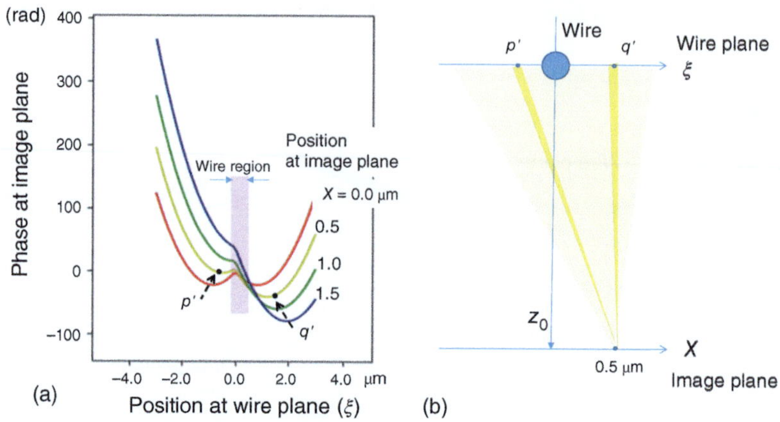

Figure 12.7 Principle of interference fringe formation with two-dimensional incident wave. (a) Phase change of wave as function of wire plane position along ξ-axis. Colored lines correspond to different positions in image plane (hologram). (b) Enlarged illustration of relationships between stationary regions and point on image plane corresponding to $X = 0.5\ \mu m$ (indicated by thick yellow lines).

incident wave is thus given by

$$\psi^*(X, Y)\psi(X, Y) = \left| \int_{-\eta_{max}}^{\eta_{max}} \left[\int_{-\xi_{max}}^{-a} + \int_{a}^{\xi_{max}} \right] \exp\{i[\phi(\xi, \eta) + \phi_U(\xi, \eta)]\}\, d\xi d\eta \right|^2 \tag{12.37}$$

Figure 12.7 shows the principle of interference fringe formation with a two-dimensional incident wave. The interference effect is caused by stable phases in two stationary regions (p' and q') on both sides of the biprism wire along the ξ-axis, as shown in Figure 12.7a. The positions and widths of the stationary regions in the wire plane depend on the parameter where the positions in the image plane are indicated by colored lines. The thick yellow lines in (b) indicate the main contribution of the two stationary regions in the wire plane to position $X = 0.5\ \mu m$ in the image plane. When the phases of the two yellow lines at $X = 0.5\ \mu m$ are equal, a strong interference effect occurs.

When a spinning linear wave passes through the wire plane from p' to q' and maintains its waveform, there is a high probability of electron propagation to position $X = 0.5\ \mu m$ in the image plane. Therefore, as in the interpretation of diffraction intensities, the degree of waveform maintenance for a spinning linear wave is assumed to be equal to the magnitude of the interference effect of the two-dimensional incident wave in the region of the line trace (l_1) along the translational motion of the spinning linear wave.

The formulation for evaluating the interference effect is given as follows. The scattering amplitude of the spinning linear wave for the kth incidence is given by

$$\psi_k(X, Y, t) = D \exp(-i\omega t_k) \iint_{l_k} \exp\{i[\phi(\xi, \eta) + \phi_U(\xi, \eta)]\}\, d\xi d\eta \tag{12.38}$$

where $\iint_{l_k} \ldots$ is the domain of integration corresponding to the area of line trace l_k on the wire plane except for the wire region. The relative intensity distribution on the X–Y image plane averaged for n electrons is given by

$$I(X,Y) = \frac{1}{n}\sum_{h=1}^{n}\psi_h^*(X,Y,t)\cdot\sum_{k=1}^{n}\psi_k(X,Y,t) \tag{12.39}$$

By neglecting the summation of the products except $h = k$, as for Eq. (12.33), and setting n to an infinitely large number, we can write Eq. (12.39) as

$$\begin{aligned} I(X,Y) &= \frac{1}{n}\sum_{k=1}^{n}\overline{\iint_{l_k}\exp\{i[\phi(\xi,n)+\phi_U(\xi,n)]\}d\xi dn} \\ &\quad\cdot\iint_{l_k}\exp\{i[\phi(\xi,n)+\phi_U(\xi,n)]\}d\xi dn \\ &\approx \left|\iint_{S_w}\exp\{i[\phi(\xi,n)+\phi_U(\xi,n)]\}d\xi dn\right|^2 \end{aligned} \tag{12.40}$$

where S_w is the area of the wire plane except for the wire region and thus the equation is equivalent to Eq. (12.37) when n is sufficiently large.

12.5.3 Simulation of Interference Fringes

Figure 12.8 shows interference fringes with spinning linear waves simulated in accordance with the formulations presented in the above section. The line traces of spinning linear waves for $n = 20$ incident electrons striking the wire plane are plotted in Figure 12.8a. The 20 electrons in the image plane were plotted along the line traces of the spinning linear waves in accordance with Eq. (12.39), using random numbers to estimate the probability of finding each electron. Figure 12.8b–d shows simulated interference fringes for $n = 2 \times 10^2$, 2×10^3, and 2×10^4 electrons, respectively. Interference fringes gradually formed as the number of incident electrons increased, as observed experimentally [25] (see Figure 5.23). A relatively smooth intensity profile (e) was obtained by integrating the intensities in (d) along the y-axis. This agrees well with the interference fringe profile obtained with the plane wave incidence (f).

12.6 Interpretation of Wave–Particle Dualism

The formulations and simulation with the spinning linear wave model described in Section 12.5 are summarized as follows. To understand both the point-charge behavior of an electron and its coherent length at the micrometer scale, the interference effect due to the incident plane wave is considered to result from many spinning linear waves. Each wave corresponds to an incident electron. The shape of a linear wave is closely related to the gravitational field of an electron corresponding to the one-dimensional density of the matter, as described by the gravitational field equations of Infeld and Schild.

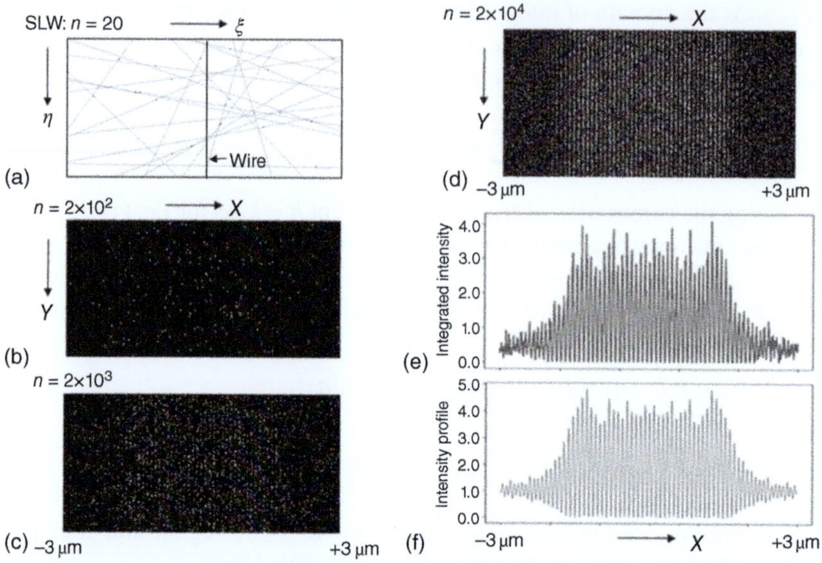

Figure 12.8 Simulated intensity profiles of interference fringes. (a) Line traces of 20 spinning linear waves on wire plane (small dots represent positions of electrons on image plane). (b–d) show simulated interference fringes for 2×10^2, 2×10^3, and 2×10^4 electrons, respectively. (e) Normalized intensity profile of interference fringes in (d) integrated along Y-direction. (f) Simulated intensity profile of interference fringes (equal to Figure 5.22d) for comparison.

The spinning motion of the linear wave is attributed to the quantized spin angular momentum with a magnitude of $h/(4\pi)$. The quick spinning motion results in high-density formation of a linear wave around its center position showing the particle picture, while the long length of the linear wave (L_{LW}) corresponds to the coherence length contributing to the interference effect.

The formulation of the electron diffraction pattern is equivalent to those obtained using the plane wave incidence when the number of incident electrons is sufficiently large. Furthermore, the process of forming interference fringes in electron holography has been successfully reproduced by simulation with spinning linear waves as a function of the number of incident electrons. The results are in good agreement with the experimental ones.

From a practical perspective, the use of the spinning linear wave model is more effective for systematically understanding the behavior of an electron in various electronic systems than the use of conventional treatments using dualistic particle and wave pictures. For example, the electron trajectory through the various electron lens systems in an electron microscope system, such as that depicted in Figure 5.5, has been generally analyzed and designed using a simple point-charge picture as a conventional treatment, as explained in Section 5.1. When the interaction of electrons with a material resulting in diffraction patterns is analyzed, the electrons have generally been considered to form a plane wave, as defined in Eq. (12.3). In contrast, the spinning linear wave model can be applied to both cases uniformly [26].

For the electron trajectory, the time-averaged highest density region of the charge in a spinning linear wave forms at a particle position, so its treatment is approximately equal to that with a point charge. The interaction of electrons with a material resulting in diffraction patterns can be analyzed as the crossings of many spinning linear waves across the material, and its evaluation is equivalent to quantum mechanical treatment with plane wave incidence in the area when the number of incident electrons is sufficiently large, as discussed in Section 12.5.1. Furthermore, when a spin-polarized electron emitter for a transmission electron microscope is available in the near future, we will be able to analyze the distribution of spin polarized electrons and their interactions with magnetic materials in detail with the present spinning linear wave model.

References

1 de Broglie, L. (1923). Ondes et Quanta. *Comptes Rendus* 177: 507.

2 G. Lochak, De Broglie's initial conception of de Broglie waves, in *The Wave-Particle Dualism*, Springer Link, ed.by S. Diner, D. Fargue, G. Lochak, and F. Selleri, D. Reidel, 1984, 1.

3 Einstein, A. and Infeld, L. (1978). *The Evolution of Physics*, 2e. Cambridge: Cambridge University Press.

4 Shindo, D., Jung Kim, J., Hyun Kim, K. et al. (2009). *J. Phys. Soc. Jpn.* 78: 104802.

5 Schlosshauer, M. (2005). *Rev. Mod. Phys.* 76: 1267.

6 Yoshioka, H. (1957). *J. Phys. Soc. Japan* 12: 618.

7 Braginsky, V.B., Vorontsov, Y.I., Khalili, F.Y., and Zh.Eksp (1977). *Teor. Fiz.* 73: 1340.

8 Braginsky, V.B., Vorontsov, Y.I., and Thorne, K.S. (1980). *Science* 209: 547.

9 Grangier, P., Levenson, J.A., and Poizat, J.-P. (1998). *Nature* 396: 537.

10 Renninger, M. (1960). *Z. Phys.* 158: 417.

11 Shindo, D. and Akase, Z. (2020). *Mater. Sci. Eng. R* 142: 100564.

12 Jackson, J.D. (1962). *Classical Electrodynamics*. New York: Wiley.

13 Einstein, A. (1916). *Ann. der Phys. Ser. 4* 49: 769.

14 Møller, C. (1952). *The Theory of Relativity*. Oxford: Clarendon Press.

15 Pauli, W. (1958). *Theory of Relativity*. London: Pergamon Press.

16 Einstein, A. (1905). *Zur Elecktrodynamik bewegter Körper. Ann. der Phys.* 17: 891.

17 Einstein, A. (1915). *Zur allgemeinen Relativitätstheorie*, 777. S.B. Press. Akad. Wiss.

18 Einstein, A. (1919). *Spielen Gravitationsfelder im Aufbau der materierllen Elementareilen ein wesentliche Rolle? Berlin Berichte* 349.

19 Einstein, A. (1969). On the generalized theory of gravitation. In: *Scientific American Resource Library: Reading in the Physical Sciences*, vol. 1, 61. W.H. Freeman.

20 Einstein, A., Infeld, L., and Hoffmann, B. (1938). Gravitation equations and the problem of motion. *Ann. Math.* 39: 65.

21 Infeld, L. and Schild, A. (1949). *Rev. Mod. Phys.* 21: 408.

22 Cowley, J.M. (1975). *Diffraction Physics*. New York: Elsevier.
23 Shindo, D. and Hiraga, K. (1998). *High Resolution Electron Microscopy for Materials Science*. Tokyo: Springer.
24 Komrska, J. (1971). *Adv. Electron. Electron Phys.* 30: 139.
25 Tonomura, A., Endo, J., Matsuda, T. et al. (1998). *Am. J. Phys.* 57: 117.
26 Shindo, D. and Tomita, T. (2020). Analyzing system with minimum electron dose. Electron Microscope Analysis System, Patent application, No. 2020–105201.

A

Physical Constants, Conversion Factors, and Electron Wavelength

Physical Constants

		SI Units	CGS Units
Elementary electric charge (e)	= 1.602 2 = 4.803 2	$\times 10^{-19}$ C	$\times 10^{-20}$ emu $\times 10^{-10}$ esu
Electron mass (m)	= 9.109 4	$\times 10^{-31}$ kg	$\times 10^{-28}$ g
Proton mass (m_p)	= 1.672 6	$\times 10^{-27}$ kg	$\times 10^{-24}$ g
Neutron mass (m_n)	= 1.674 9	$\times 10^{-27}$ kg	$\times 10^{-24}$ g
Velocity of light (c)	= 2.997 9	$\times 10^{8}$ ms^{-1}	$\times 10^{10}$ c m s^{-1}
Mass energy of an electron (mc^2)	= 8.187 1 (= 0.511 00 MeV)	$\times 10^{-14}$ J	$\times 10^{-7}$ erg
Planck's constant (h)	= 6.626 1	$\times 10^{-34}$ Js	$\times 10^{-27}$ erg s
$\hbar = h/2\pi$	= 1.054 6	$\times 10^{-34}$ Js	$\times 10^{-27}$ erg s
Compton wavelength ($\lambda c = h/mc$)	= 2.426 3	$\times 10^{-12}$ m	$\times 10^{-10}$ cm
Avogadro's number (N_A)	= 6.022 1	$\times 10^{23}$ mol^{-1}	$\times 10^{23}$ mol^{-1}
Bohr magneton (μ_B)	= 9.274 0	$\times 10^{-24}$ J T^{-1}	$\times 10^{-21}$ erg Oe^{-1}

Conversion Factors

1 eV = 1.6022 × 10^{-19} J	1 Å = 0.1 nm 1 Oe = 79.577 AT cm^{-1}
Torr = 133.32 Pa	1 G = 10^{-4} T = 0.79577 AT m^{-1}

Material Characterization using Electron Holography, First Edition. Daisuke Shindo and Takeshi Tomita.

Electron Wavelength and Interaction Constant

Accelerating Voltage, V (kV)	Wavelength, λ (nm)	Interaction Constant, σ (V^{-1} nm^{-1})
80	0.004 175 72	0.010 087 1
100	0.003 701 44	0.009 244 0
120	0.003 349 22	0.008 638 1
150	0.002 957 04	0.007 989 2
180	0.002 665 50	0.007 528 4
200	0.002 507 93	0.007 288 4
300	0.001 968 75	0.006 526 2
400	0.001 643 94	0.006 121 4
500	0.001 421 26	0.005 873 2
600	0.001 256 80	0.005 707 2
700	0.001 129 28	0.005 589 7
800	0.001 026 95	0.005 503 0
900	0.000 942 69	0.005 436 8
1000	0.000 871 92	0.005 385 0
1250	0.000 735 71	0.005 295 6
1300	0.000 713 61	0.005 282 4
1500	0.000 637 45	0.005 239 7
2000	0.000 504 32	0.005 176 0
2500	0.000 417 83	0.005 142 3
3000	0.000 356 93	0.005 122 3

Index

Material Characterization using Electron Holography, First Edition. Daisuke Shindo and Takeshi Tomita.

f